THE DYNAMICS OF
SPECTRUM
MANAGEMENT

THE DYNAMICS OF
SPECTRUM
MANAGEMENT

Legacy, Technology, and Economics

ROHIT PRASAD
V. SRIDHAR

OXFORD
UNIVERSITY PRESS

OXFORD
UNIVERSITY PRESS

Oxford University Press is a department of the University of Oxford.
It furthers the University's objective of excellence in research, scholarship,
and education by publishing worldwide. Oxford is a registered trademark of
Oxford University Press in the UK and in certain other countries

Published in India by
Oxford University Press
YMCA Library Building, 1 Jai Singh Road, New Delhi 110 001, India

© Oxford University Press 2014

The moral rights of the authors have been asserted

First Edition published in 2014

ISBN-13: 978-0-19-809978-9
ISBN-10: 0-19-809978-9

Typeset in Adobe Garamond Pro 11/13
by SPEX Infotech, Puducherry, India 605 009
Printed in India by Sapra Brothers, New Delhi 110 092

To my thesis advisor, Pradeep Dubey,
who compelled me to think, and to my parents,
Pandey Surendra Prasad and Meera Prasad,
who inspire me to act.

—Rohit Prasad

To my late father, K.S. Varadharajan,
for his passion towards engineering and physics;
and to my beloved wife, Kala, and daughter,
Vindhya, for their enduring love and support.

—V. Sridhar

Contents

List of Figures and Tables   xi
Preface   xvii
Acknowledgements   xix

1 History and Technology of Spectrum   1
       What are Radio Waves? 1
       The Birth of Radio 3
       The Problem of Interference 4
       Spectrum Allocation Process: International and National 5
       Early Days of Wireless Mobile Systems 16
       2G Mobile Systems 22
       Third Generation Mobile Technologies 27
       Fourth Generation Wireless Networks and beyond 29
       Wi-Fi and the Spectrum of Commons 37
       Future of Wireless Technologies 40
       Conclusion 45
       Annexure A1 46
       References 49

2 The Economic Classification of Spectrum and
       the Commons Debate   52
       The Economic Classification of Goods 53
       The Economic Classification of Spectrum 55
       The Tragedy of the Commons 57
       Resolving the Tragedy of the Commons 58
       Capacity-enhancing Technological Progress 60
       Community Solutions 62
       The Debate over the Spectrum Commons 64
       A Middle View 68

Conclusion   69
References   70

3   Conceptual Issues in the Allocation and
Assignment of Spectrum   71
Introduction   71
Determining Spectrum Use   73
Choosing Standards   74
Number of Blocks   78
Setting the Spectrum Price   81
Conclusion   83
References   83

4   Auctions   85
Introduction   85
Auction Formats   87
A Brief History of Spectrum Auctions   89
Modelling Auctions   90
Findings from Auction Theory   91
Simultaneous Ascending Auctions   96
Package Auctions   98
Conclusion   102
References   102

5   Economic Models for Valuing Spectrum   103
Introduction   103
*Estimating the Value of Spectrum
Using the Cash Flow Method*   105
*Estimating the Value of Spectrum
Using the Production Function Method*   111
Conclusion   116
References   117

6   Spectrum Transitions: Moving from Command and
Control to Flexible Use   118
Introduction   118
Command and Control versus Flexible Use   119
Rationalization of Government Spectrum   121
New Licensing Framework   126
Creating Secondary Markets   127

Flexible Use   138
Smaller and More Frequent Auctions   141
Universal Service Obligation   141
Conclusion   142
References   143

7 Pathways to the Spectrum Commons   144
Introduction   144
History   144
Spectrum Sharing Frameworks   147
Wi-Fi   148
Spectrum Sharing   153
Dynamic Opportunistic Spectrum Access   155
Case Studies of Unlicensed Use   159
Conclusion   161
References   163

8 Net Neutrality   164
Overview   164
Congestion in Networks   168
The Market Power of the ECP   173
Vertical Integration in the ICT Industry   180
Net Neutrality in Practice   184
Implications for Spectrum Regulation   185
Conclusions   186
References   188

9 Mobile Partnerships and Alliances   191
Introduction   191
Command and Control   191
Flexible Spectrum Management   199
Commons   207
Partnerships across the Telecom Value Chain   212
Conclusion   215
References   218

10 Hyper Competition and Excessive Spectrum
    Fragmentation: Case of India   219
Introduction   219
The First Stage: 1995–2003   222

*Unified Access Service Licence and
Administrative Pricing of Spectrum: 2003–8*   230
Rush to Acquire Mobile Licence and Spectrum Post 2007   233
Liberalization of Spectrum 2008 Onwards   239
The Possibilities of Cognitive Radio   289
Spectrum for Research and Development   291
Spectrum for Development   293
Conclusion   294
References   296

11 Spectrum for Broadcasting Services   299
Introduction   299
Radio Broadcasting   300
Television Broadcasting   307
Future Directions   321
Conclusions   323
References   323

12 The Way Forward   325
Legacies   326
Technologies   327
Economics   328
Conclusion   328
References   329

Glossary   331
Bibliography   347
Index   357
About the Authors   367

Figures and Tables

Figures

1.1	Cabin of an amateur radio operator	5
1.2	Wireless apparatus on board *Titanic*, Marconi, and Jagadish Chandra Bose	8
1.3	ITU regions map	11
1.4	US frequency allocation chart	12
1.5	Nomenclature of various frequency bands	15
1.6	A typical cell pattern for re-use in cellular networks	19
1.7	Evolution of 3G technologies	28
1.8	Evolution path for 4G	31
1.9	Mobility and user data offerings of different technologies	34
1.10	Small cells dense architecture	43
1.11	Future network architecture	44
6.1	Femtocell architecture	136
7.1	Architecture of the carrier Wi-Fi network elements	151
7.2	Architecture of the Mi-Fi networks	153
8.1	End-user experience at different levels of capacity	167
8.2	Global IP traffic developments (fixed and mobile)	168
8.3	Mobile data traffic trends	169
8.4	Fixed and mobile subscriptions, 2008–16	170
8.5	Growth of mobile subscriptions vis-à-vis landline subscriptions in India	171
8.6	Market structure of the Internet	175
8.7	Illustration of the two-sided markets	175
9.1	The enhanced mobile value chain	214
10.1	Allocation of spectrum between ministries of defence, and communication and IT in India	221
10.2	Subscriber based criterion for spectrum assignment	229

10.3	Growth in subscriber base post 2008 along with key reasons	236
10.4	Decline in ARPUs post 2008	237
10.5	Carrier distribution of GSM spectrum allotted to telecom service providers licence area network wise	274
10.6	Non-contiguity of spectrum assignment across circles to various operators in the auction in February 2014	285
10.7	Indian ER&D offshoring market	291
10.8	Types of licence issued for ERD and testing	292
11.1	Spectrum used in radio and television broadcasting	300
11.2	Amount of spectrum that could be vacated by switchover from analog to digital	312
11.3	Harmonized FDD and TDD arrangement in 698–806 MHz band	314
11.4	600 MHz band plan in the proposed incentive auction in the US	318

Tables

1.1	Commonly used radio frequency list	10
1.2	Usage designations of radio frequencies in India	13
1.3	Letter designations of radio frequencies	15
1.4	Evolution of 2G technologies	27
1.5	Comparison of various mobile technologies	32
1.6	Deployment of 3G and 4G technologies across different spectrum bands in different countries	35
1.7	The evolution of IEEE 802.11 Wi-Fi access network technologies	39
2.1	The economic classification of goods	54
2.2	Overcrowding on a common pasture	58
2.3	Overcrowding on a common pasture despite enhanced capacity	61
3.1	Unequal values in different uses	73
3.2	Equal values in different uses after re-allocation	73
3.3	Exclusive use of spectrum band for single use	74
4.1	Auction formats	88
5.1	Cost of BTS	107
5.2	Annual spectrum charges	108

5.3	Licence fees for different categories of circles	108
5.4	Calculation of value of contracted spectrum in Tamil Nadu	110
5.5	Value of spectrum using the cash flow method	110
5.6	Estimated parameters of the production function	114
5.7	Value of spectrum using the production function method	115
6.1	Difference between command and control and flexible use regimes	120
6.2	Spectrum allocation for commercial services in India and Finland	122
6.3	Examples of partnerships in spectrum trading	128
6.4	Disparity of spectrum allocation for wireless broadband services in India	132
6.5	Types of transactions in spectrum in secondary markets	138
7.1	Frameworks for non-exclusive use	148
7.2	Frameworks for exclusive and non-exclusive use	162
9.1	Recent examples of inter-region acquisitions	193
9.2	Controversial inter-region acquisitions in India	194
9.3	Major partnerships for tower sharing	199
9.4	The 3G roaming pact	202
9.5	Examples of significant MVNOs in the world	205
9.6	Countries that have a large number of MVNOs	206
9.7	Partnerships in Femtocell deployments	207
9.8	Partners of British Telecom for Wi-Fi access in countries around the world	209
9.9	Most notable free Wi-Fi offered by cities	211
9.10	Examples of partnerships across the value chain	216
10.1	Circles for mobile services	223
10.2	Licence fee commitments for cellular mobile services	224
10.3	Spectrum allocation criterion in 2002	230
10.4	Spectrum allocation criterion for additional spectrum allocation in 2008	231
10.5	2008 policy outcomes	235
10.6	Airtel valuation ratios	238
10.7	Companies whose licences were cancelled	240
10.8	The 3G/BWA auction process parameters	244
10.9	Financials of the 3G spectrum auction	245

10.10   Financials of BWA spectrum auction   247
10.11   Timeline for the expiry of licence and
        associated spectrum   253
10.12   Spectrum usage charges   255
10.13   Availability of spectrum for auction in the 1800 MHz
        and 800 MHz bands   258
10.14   Reserve prices for the various bands recommended
        by TRAI   260
10.15   Reserve prices for various spectrum blocks auctioned
        during 2012–13   262
10.16   Results of the 1800 MHz spectrum auction held in
        November 2012   263
10.17   Results of the 800 MHz spectrum auction held in
        March 2013   268
10.18   The status of spectrum allocation post March 2013   271
10.19   Status of spectrum assignment across LSAs post
        March 2013 auction   272
10.20   Results of the 1800 MHz spectrum auction held in
        February 2014   276
10.21   Results of the 900 MHz spectrum auction held in
        February 2014   279
10.22   Average spectrum assignment per operator in different
        LSAs as on July 2013   282
10.23   Comparison of spectrum prices across different LSAs
        for different bands   287
10.24   Comparison of spectrum prices across different
        countries and that of India   288
10.25   Important milestones in the Indian mobile sector   294
11.1   Categorization of cities for FM broadcasting service   302
11.2   Radio frequency allocation for FM broadcasting in
        NFAP 2011   303
11.3   Details of the FM auction   304
11.4   Allocation of FM radio channels in each city category
        in Phase III allocation in India   307
11.5   Models of TV broadcasts   308
11.6   Frequencies and number of channels provided
        by Doordarshan   310

11.7 Progress on digital switchover of terrestrial TV in
      select countries   313
11.8 Clauses in NFAP 2012 regarding digital
      dividend spectrum   313
11.9 INSAT satellites for DTH services   320
11.10 Timelines of various policy and regulatory decisions
      in the television sector   322

Preface

Telecommunications, along with a few other sectors such as energy and healthcare, represents a fascinating confluence of high technology, developmental impact, and economic transformation. The role of wireless communications is especially important given the ubiquity of the mobile phone and the rapidly growing uptake of mobile broadband.

Spectrum, an essential input in mobile services, is a critical subject for study, as policymakers grapple with the dilemmas of access and empowerment, service providers gear up for bandwidth overload, and technologists strive to unlock new modes of maximizing resource productivity. The regulatory issues of economies of scale, provision of universal service, and technological dynamism that characterize telecommunications as a whole are inherited by its inputs, including spectrum. However, spectrum presents unique challenges of its own that play a defining role in the industry. The phenomenon of economies of scale of the mobile industry is driven in large part by the 'trunking efficiency' characterizing spectrum use. The nature of partnerships seen stems from strategic imperatives related to spectrum. Even more complex conundrums are presented by the prospect of the simultaneous use of spectrum bands using dynamic spectrum access modes and spread spectrum techniques that potentially change the nature of spectrum from a private good to a commons.

While further technological clarification, institutional innovation, and policy enablement will determine the precise contours of future spectrum management, current spectrum management regimes across the world exist in widely disparate stages of sophistication. Some are in the basic mode of *command and control*, where the government decides the type of use, terms of payment, and tenure of spectrum use. Others have moved to *market-oriented* mechanisms to varying degrees. Historically, there has also been a practice of unlicensed spectrum use which continues to this day, and is buttressed by modern techniques of spectrum sharing. Each jurisdiction has to move ahead by carefully balancing legacy constraints, technological possibilities, and economic imperatives.

This book aims to present the transition paths that each jurisdiction can adopt, starting from their present position to frameworks that will define the future. It starts with an introduction to the technology of wireless communications, proceeds to outline the economic principles that must inform spectrum management regimes, and then presents the dynamics of the transition paths from legacy regimes. Finally, the case study of India for wireless communications and broadcasting is presented.

In the spirit of an economic resource that has the potential of changing the destiny of large masses of the world's population, the writing style favours simplicity over technical jargon, and accessibility over abstruseness, hopefully without sacrificing substance or depth. Awareness about the immense potential inherent in the wise use of spectrum among large sections of the population can have a far-reaching impact. We look forward to a day when the right to airwaves will be as pervasive as the right to air.

Acknowledgements

This book is a product of our unique privilege of interacting with the best minds in academia, policymaking, and the ICT industry. Experiences at the Department of Telecommunications (DoT), Government of India, in the Subodh Kumar Committee on 2G spectrum during 2008–9; in the Telecommunications Regulatory Authority of India (TRAI) Expert Panel on Spectrum Pricing during the tenure of J.S. Sarma; expert testimony for mobile operators; and interactions with national and international experts in the area of telecommunications have informed and enriched our understanding of the topic. We are especially grateful to D. Manjunath of the Indian Institute of Technology Bombay; Rakesh Mehrotra of the AKG Engineering College; Sajal Ghosh of Management Development Institute (MDI), Gurgaon; Rupamanjari Sinha Ray of MDI, Gurgaon; Parag Kar of Qualcomm; P.K. Garg, ex-wireless advisor to the Government of India; Rajat Kathuria of the Indian Council for Research on International Economic Relations (ICRIER); Heikki Hämmäinen, Thomas Casey, and Arturo Basaure of Aalto University, Finland, for their insights and valuable advice that have been incorporated throughout the book.

Thanks to our colleagues at Sasken and MDI, Gurgaon, for the many discussions on related topics and for providing the intellectual inputs that have been incorporated in some form or the other in the book. They have consistently shown remarkable understanding and empathy for our pre-occupation with our research interests. To them we owe our heartfelt gratitude.

Our mentors in academics, Pradeep Dubey of Stonybrook and Yale universities and June Park at the Korea Advanced Institute of Science and Technology, Seoul, have provided us the best foundation on which to build a life—an appreciation for the value of scholarship, and an aspiration to strive for rigour in all we do.

A book is the product of a large team of people. We received valuable research assistance and support from Geetanjali Vijayran. Her artistic flair is demonstrated in the figures and tables throughout the book. Special thanks to Chakravarthy Buchi of Sasken Communication Technologies for reviewing the first chapter of the book and providing great tips for improvement. Thanks to Subhash Rao of Sasken Communication Technologies that we were able to bring the archival photo of Jagadish Chandra Bose to a reality sketch.

The two anonymous reviewers of our manuscript, through their critical remarks and insights, helped us to rewrite and improve many parts of the book; thanks to both of them for their commitment to the review process. Thanks to the editorial team at the Oxford University Press which diligently worked with us for the past one year to make the book what it is today.

Finally, our families are the wind beneath our wings, our reason to step out and do our best. Sonali, Chaitanya, and Soham; Kala and Vindhya—this book is for you!

1

History and Technology of Spectrum

In the new era, thought itself will be transmitted by radio.
—Guglielmo Marconi, Inventor of Radio

What are Radio Waves?

Radio, television, and cellular mobile phones have one thing in common: 'communicating over the air'. The communication 'airwaves', also referred to as the 'radio frequency spectrum', are an essential resource in modern communication because there are more cellular mobile phones than traditional wired telephones in the world today, more satellite TV sets than cable-connected TVs, and many consumer devices, including mobile phones, have radio antennas to provide FM Radio services.

The radio frequency spectrum refers to a subset of frequencies of electromagnetic spectrum. Electromagnetic spectrum is the range of all possible frequencies of electromagnetic radiation, which in turn is a form of energy emitted and absorbed by charged particles as they travel through space exhibiting wave-like behaviour. The electromagnetic spectrum extends from below the low frequencies used for modern radio communication to gamma radiation at the high frequency end. Radio spectrum refers to the part of the electromagnetic spectrum that can be used for communication. It corresponds to frequencies from 3 KHz to around 300 GHz. Above 300 GHz, the absorption of electromagnetic radiation by the earth's atmosphere is so great that the atmosphere is effectively opaque, until it becomes transparent again in the near-infrared and optical window frequency ranges.

Radio waves inherit the essential properties of waves. They are characterized by their wavelength and their frequency, which bear an inverse relationship with one another. Electromagnetic waves travel at the speed of light, 3×10^8 m/sec. A radio channel which transmits at a frequency of 1000 KHz (1 MHz or 1 million waves per second) will have a wavelength of 300 m.[1] If it were to transmit at three times the frequency, that is, at 3000 KHz, it will have a wavelength one-third as large, that is, 100 m. The inverse relationship between frequency and wavelength has important implications as it accords greater propagation capability to lower frequency spectrum.

While low frequency spectrum is preferred on account of its propagation characteristics, a high frequency spectrum is more desirable in terms of its ability to squeeze in a greater amount of information in each frequency band, a property measured by bandwidth.

The power of transmission is another variable that determines the coherence of a particular signal and the distance to which it is propagated. For instance, some of the AM Radio stations originating in countries such as the US, the UK, Russia, and Germany have listeners around the world.

The ability to transmit signals using radio waves has made devices using spectrum ubiquitous in daily life. The applications include AM Radio and FM Radio, wildlife tracking collars, garage door openers, terrestrial and satellite TVs, mobile phones, cordless phones, microwave ovens, deep space radio communications, and even baby monitors. We turn to a brief account of the people and the events which led to the birth of radio.

---

[1] Velocity of propagation of radio waves (v) in metres/second = 3×10^8 = wavelength λ (in metres) × frequency f (in hertz or number of oscillations of the wave per second).

The Birth of Radio

Heinrich Hertz's historic experiments dealt with generation and reception of electromagnetic waves of wavelength 30 cm to 8 m. However, he could not conceive of the idea of using radio waves for voice communications. In 1889, he stated that continent- size dishes would be necessary to send audio frequency range wireless waves for voice communications. It was Guglielmo Marconi who conceived the idea of 'modulating' electromagnetic waves using a Morse code of dots and dashes to send information. This implied generating a short duration wave for a dot, a longer wave for the dash, and no wave for a space. In the face of conventional wisdom, Marconi also believed that it would be possible to send information riding on electromagnetic waves over the horizon. Other scientists disputed this, since they believed that electromagnetic waves travelling in a straight line would diverge off the surface of the earth which is spherical. To test his hypothesis, Marconi established a huge transmitting antenna at Poldhu, Cornwall, UK.

On 12 December 1901, Marconi received the letter 'S' (three dots) at Newfoundland, St. Johns, was transmitted across the Atlantic Ocean over a distance of 2,000 miles from Poldhu. The receiving antenna consisting of a 150 m long wire was put on a kite, from which a wire came to a pole and finally to the receiver. The receiver consisted of a 'coherer' and a telephone receiver to hear the signal using an earphone. This was Macroni's famous transatlantic experiment which surprised the world (Aggarwal 2006).

Though Guglielmo Marconi of Italy is credited with the invention of sending and receiving radio signals over long distances, the contributions of India-born Sir Jagadish Chandra Bose need special mention. In a salutation to Sir Bose, Varun Aggarwal, a PhD student at MIT, USA, maintains a web page citing his early contributions (Aggarwal 2013). Aggarwal shows that the Mercury *autocoherer* in the receiving device used by Marconi was in fact invented by Sir Bose in 1899 (Aggarwal 2006). The coherer that is connected to the antenna, converted the AC (RF signal) to a DC signal (or a low frequency signal), which could then drive a Morse printer or an earphone. The autocoherer returned to its initial state automatically after the signal was received (Aggarwal 2006). The details of this invention were presented by Sir Bose at The Royal Society and published in his classic paper (Bose 1899). However, it was Marconi who patented the steel–mercury–carbon coherer in 1901 in his name [see Aggarwal (2013) for some interesting facts on the history of telecommunications].

The Problem of Interference

One of the most important characteristics of spectrum is interference. To understand interference, imagine dropping a pebble into a still pond. The waves generated by the pebble are clearly visible. However, if 20 pebbles are dropped simultaneously, the waves from the pebble will no longer be visible. In a similar manner, the transmission of signals of sufficiently high power by multiple users at the same time and at the same frequency in close physical proximity leads to distortion that can inhibit or entirely block communication.

Counter-intuitively, unlike ripples in a lake, electromagnetic waves are not destroyed on impact but pass through each other. Hence the use of superior listening devices that can focus on the relevant signal can allow communication to continue even in the presence of multiple signals being sent in close proximity. This is akin to conversations being carried out in a crowded, noisy room where each listener focuses on the sounds emanating from his or her circle.

In the early days of spectrum use, interference was a relatively easy challenge to handle as there was a lot of spectrum and limited usage. The freedom to use spectrum led to the growth of amateur radio, the use of designated radio frequency spectrum for purposes of private recreation, non-commercial exchange of messages, wireless experimentation, self-training, and emergency communication. Today, amateur radio is still used by hobbyists and at the end of 2011, there were an estimated 2 million amateur radio users in the world. Figure 1.1 gives an idea of an amateur radio cabin.

As the use of radio spectrum increased, interference became an issue and led to the advent of spectrum regulation. The problem of interference cannot be better explained than by the tragedy of *Titanic* as explained in Exhibit 1.1 (Stephenson 2002).

After the *Titanic* disaster, the Radio Act of 1912 was authorized in the US by the Secretary of Commerce to licence users of equipment

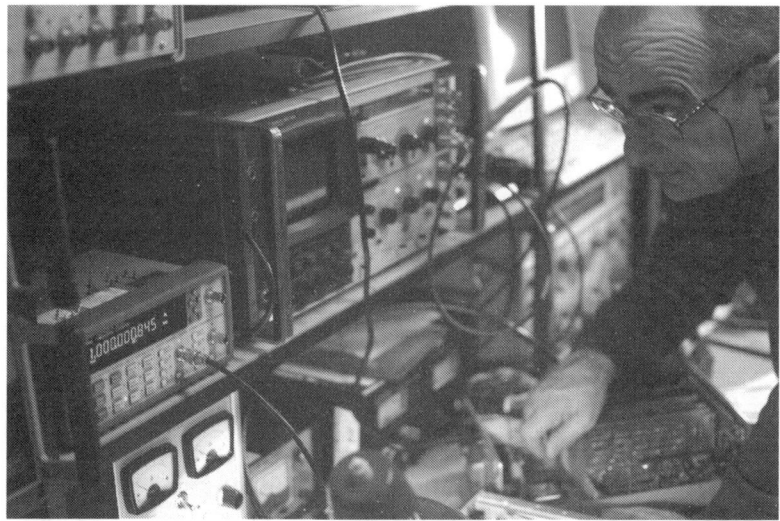

Figure 1.1   Cabin of an amateur radio operator

that communicated via spectrum (Nuechterlein and Weiser 2005). The act precluded interference by restricting access to spectrum. With the enactment of the Radio Act in 1927, the Federal Radio Commission was set up in the US to regulate access to spectrum under a general 'public interest' standard.

Spectrum Allocation Process: International and National

Radio signals do not stop at national boundaries. Sometimes a specific region in a country may have a number of international borders (for example, borders of Jammu and Kashmir in India). Hence, international regulation is necessary to address the issue of interference. International coordination of the use of spectrum is also occasioned by the scarcity of spectrum bands, the necessity for cross-border operability of communication devices, and the economies of scale associated with spectrum and devices.

The organization tasked with coordinating spectrum use at a global level is the International Telecommunications Union (ITU). The ITU was founded in Paris in 1865 as the International

Exhibit 1.1   The *Titanic* story: An example of radio interference

In the early morning hours of 15 April 1912, a high-pitched musical tone sang out for hundreds of miles across the North Atlantic in a desperate plea for help. The White Starliner *R.M.S. Titanic* had struck an iceberg, and her 5 kW Marconi installation was signalling her death knell. Over the years, *Titanic*'s Marconi wireless set and her steadfast operators have attracted a good deal of public attention. *Titanic* as well as wireless historians have offered technical descriptions of the equipment and its capabilities in various publications throughout the years. *Titanic*'s wireless set had a nominal working range of 250 nautical miles, but signalling more distant stations was possible. At night, ranges of up to 2,000 miles were attained with sets of similar architecture. On board *Titanic*, the wireless equipment was housed in a series of inter-connecting rooms—the sound-proof Silent Room in which noisy transmitting equipment was located; the Marconi Room, an office which contained the operators' work stations, manipulation keys, and receiving equipment; and the Bed Room, which contained the operators' berthing.

Two employees of Marconi, the company that made the system, operated the radio. It was the most powerful system of its kind, and the clear night helped the signals go far. The transmitter was designed to send dots and dashes over a specific frequency that accommodated lots of other ships. Receivers on that frequency, whether on other ships or land-based stations, picked up the dots and dashes to get the message. To ensure the identity of the senders, each ship had a unique call sign—the *Titanic*'s was 'MGY' (Danigelis 2012). Not very sophisticated by today's standards, wireless radio communication in those days was as popular as the Internet today. But not everyone had their own iPhone. Instead, Marconi radio operators, Jack Phillips and Harold Bride, on board the *Titanic* spent a great deal of time transmitting and receiving telegrams for the passengers on a variety of topics including political developments, family updates, and sporting news.

Constant communication from high-profile passengers kept the operators busy. Passenger communications were so pressing that

Phillips got irritated when the radio operator from the *Californian*, a ship nearby, interrupted one of his transmissions with a now infamous courtesy message saying that they had encountered ice. Phillips angrily signalled back, 'Shut up!' The incoming message was jamming him, and he was busy.

When *Titanic* hit the iceberg, its radio engineers could not afford to ignore reality any further. They sent an emergency message in the dots and dashes of the Morse Code: 'Come at once. We have struck a berg.' Many ships did receive the call. So did land-based stations in the US and Greenland. Radio operators at the time were skilled at transmitting messages quickly in code—80 to 100 words per minute. With such capabilities, what went wrong?

The problem was that the radio operator had to repeat the emergency message for each ship that he was talking to. The emergency went out as both 'CQD', a distress code used by Marconi operators, and the standard 'SOS'. Unlike modern communications where it is possible to dial a phone number and reach one person, during the *Titanic* disaster everyone was signalling at the same time. 'With the wireless, essentially it's a party line', Hayes said. Communications chaos ensued which eventually led to the sinking of *Titanic* with no help in sight. Figure 1.2 shows a glimpse of the radio apparatus on board *Titanic*, along with the two inventors of radio communications.

Source: Adapted from Stephenson (2002).

Telegraph Union. It took its present name in 1932, and in 1947 it became a specialized agency of the United Nation. Although its first area of expertise was the telegraph, ITU's work now covers the whole ICT sector, from digital broadcasting to the Internet, and from mobile technologies to 3D TV. An organization run as a public-private partnership since its inception, ITU currently has a membership of 193 countries and some 700 private sector entities. It is headquartered in Geneva, Switzerland, and has 12 regional and area offices around the world.

Radio communication services vary in their spectrum requirements, power levels of transmission, commercial viability, and

Figure 1.2 (Anticlockwise from top) Wireless apparatus on board *Titanic*, Marconi, and Jagadish Chandra Bose
Source: Wireless apparatus on board *Titanic*: http://commons.wikimedia.org/wiki/File%3ATitanic's_Marconi_appartus.jpg
Marconi: http://commons.wikimedia.org/wiki/file%3AGuglielmo_Marconi.org
Jagadish Chandra Bose: Authors' own, courtesy: Subhash Rao

economic impact. The process of demarcating certain frequencies for specific services/applications over a specific geographical area is referred to as spectrum allocation. The process of spectrum allocation is a multi-level activity presided over at a global level by the ITU. The ITU's International Radio Regulations prescribe specific uses, sometimes more than one, for different bands. The regulations also prescribe mechanisms for inter-country coordination and, in some cases, suggest ways in which certain bands are to be assigned to individual users. A key part of ITU's business is conducted through world radio conferences held every three to four years in which national spectrum managers and private stakeholders participate. The ITU does not have the power to constrain countries to manage spectrum in line with its prescriptions. However, each ITU member state is required 'to avoid causing harmful interference to services rendered by users using frequencies assigned in accordance with the International Radio Regulations' (Jakhu 2007). Table 1.1 gives a sample spectrum allocation table.

Getting multiple countries to agree to allow common or *harmonized* frequency bands for any service is not an easy task. Often, those particular frequency bands might already be in use for some other purpose and moving or shifting to other bands is a complex and time-consuming exercise. This becomes all the more difficult if the existing operations belong to a strategic user or organization such as the military or space sciences. However, due to the immense socioeconomic benefits at a national level as well as the facility of international roaming, most of the countries have adopted harmonized spectrums.

For the purpose of frequency allocations, the world has been divided into three regions (R1, R2, and R3) (see Figure 1.3) and these allocations are included in the ITU's Radio Regulations table of frequency allocations. India is in R3.

There are also certain regional bodies that coordinate the use of spectrum, such as the European Union (EU) and the Confederation of European Post and Telecommunications Agencies (CEPT). The coordination of such bodies may be more specific than that of the ITU. For instance, rather than merely allocating a band for mobile communication, they may designate it to a specific standard, such as GSM. As in the ITU's case, the recommendations of such regional bodies are, in general, non-binding. The EU, however, has legal powers to mandate member states to set aside spectrum in particular bands for specific technologies.

Every country also has a spectrum manager at a national level. The Federal Communications Commission (FCC) allocates and assigns

Table 1.1   Commonly used radio frequency list

Type of use	Frequency range
AM Radio	535 KHz–1.7 MHz
Wildlife tracking collars	120 KHz–150 KHz;
	215 MHz–220 MHz
Short Wave Radio	5.9 MHz–26.1 MHz
RFID for Smart cards	13.56 MHz
Citizens Band Radio	26.96 MHz–27.41 MHz
Garage door openers, alarm systems	~40 MHz
Baby monitors	49 MHz
Terrestrial TV	54 MHz–220 MHz
Radio controlled airplanes	~72 MHz
FM Radio	88 MHz–108 MHz
MIR space station	145 MHz–437 MHz
RFID for defence applications	433 MHz
Cell phones	450 MHz–2600 MHz
Cordless phones	~900 MHz, ~2100 MHz
Commercial RFID readers	865 MHz–868 MHz;
	902 MHz–908 MHz
Global Positioning System (GPS)	1227 MHz–1575 MHz
Air traffic control radar	960 MHz–1215 MHz
Microwave oven	~2100 MHz
Wi-Fi, Bluetooth	2400 MHz–2483.5 MHz
Deep space radio communications	2290 MHz–2300 MHz
Mobile systems	450 MHz–2600 MHz
Public Safety Wireless LAN	4940 MHz–4990 MHz
Direct to Home Satellite Television	12000 MHz–14000 MHz

Source: Brain (2000).

frequencies in the US. In India, the National Frequency Allocation Plan (NFAP) is the policy document which outlines the allocation of different parts of the frequency spectrum for various purposes. The plan is entrusted to the Wireless Planning & Coordination (WPC) group under the Minister of Communications and Information Technology (C&IT). It is revised every two years. All kinds of radio communication technologies and ministries use the spectrum for

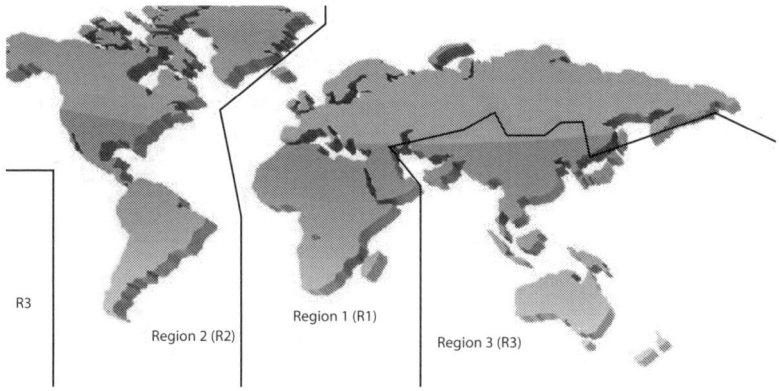

Figure 1.3   ITU regions map
Source: Retrieved from ww4.plala.or.jp/nomrax/ITU_Reg.htm on 10 January 2013.

various purposes, including space communication, radio astronomy, television broadcasting, radio navigation (for maritime uses), and, last but not the least, mobile communication.

The process of forming the NFAP is initiated by the NFAP's committee which floats a circular requesting all stakeholders to present their requests for particular blocks of spectrum. Typical stakeholders include government ministries like defence, information and broadcasting, and home affairs as well as private telecom operators and equipment manufacturers. Next, an NFAP review-revision committee invites all the stakeholders for a discussion on the allocation. This is facilitated by the formation of what are called the 'NFAP Working Groups'. Each group deals with a specific band of spectrum, such as up to 1 GHz, between 1–10 GHz, and greater than 10 GHz. Often, the discussions within in the review-revision committee lead to disagreements between different stakeholders. The irresolvable conflicts typically result in the formation of a group of ministers.

Figure 1.4 depicts the allocation of different bands of spectrum for various uses in the US and reflects the complexity of frequency allocation. Similar services are generally located in a particular band. For instance, radio broadcasting is traditionally done in lower frequencies that have superior propagation characteristics and lower attenuation across concrete obstacles. Amplitude Modulation (AM)

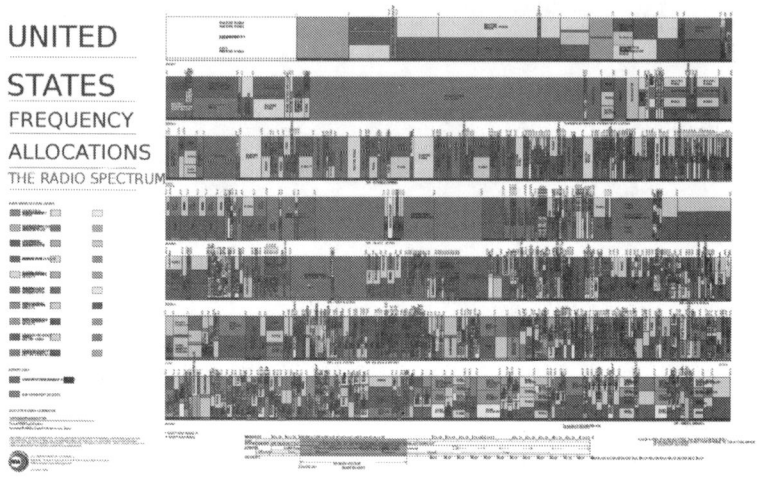

Figure 1.4   US frequency allocation chart
Source: FCC (2012).

radios use medium frequency bands in the range of 526.5 KHz to 1705 KHz. In contrast, Frequency Modulation (FM) radios use Very High Frequencies (VHF) in the range of 30 MHz to 300 MHz with propagation limited to 50–300 km, and hence are often heard only in cities. The government, police, fire and commercial voice services also use FM on special frequencies. Earlier, police radios used AM receivers to receive one-way dispatches.

Table 1.2 gives a sample of the NFAP (2011) of India.

Like other national jurisdictions in India, the Department of Telecommunications (DoT) specifies the conditions of radio activity across the borders of the country through appropriate guidelines to ensure national safety. For example, the DoT amendment order in the licence guidelines for cellular mobile telecommunication services states (DoT 2008: 2):

(1) Licensee shall ensure that the base stations, cell sites, or radio transmitters to provide mobile telephone services near the International Border of India, wherever located and established, shall be far away from such border as feasible and such equipment shall work in such a fashion that radio signals emanating therefrom,

Table 1.2 Usage designations of radio frequencies in India

Type of Use	Frequency Range
Very low power devices such as aircraft tyre pressure indicator systems on non-interference, non-protection, and non-exclusive basis	125–135 KHz
Fishing vessels	2010.4 KHz and 2025 KHz
Shipping industry	3698 KHz and 5883 KHz
Radio microphones	36.5, 36.7, 37.1, 37.9, 160.9, and 161.8 MHz
FM broadcasting	88–100 MHz; 103.8–108 MHz
Car rallies and sports activities	143.950, 150.175, and 150.9 MHz
Outside broadcast vans and film shootings	150.525, 151.250, and 166.950 MHz
Public Mobile Radio Trunked System (PMRTS)	338–430 MHz; 811–819 MHz; 856–864 MHz
Very low power remote cardiac monitoring RF wireless medical devices, medical implantation	402–405 MHz
Digital seismic telemetry	406.1–450 MHz
Digital broadcasting services including mobile TVs	585–698 MHz
RFID equipment and other low power wireless devices	865–867 MHz
Deep space research	2110–2120 MHz; 2290–2300 MHz
INSAT system for radio networking, cyclone warning dissemination system, meteorological data dissemination, satellite time frequency dissemination, and broadcast satellite service	2535–2655 MHz
Earth exploration satellite	18.6–18.8 GHz

Source: Department of Telecommunications (DoT 2011).

fade out when nearing or about to cross international border and become unusable within a reasonable distance across such border.
(2) In the areas falling within 10 km of Line of Control, installation of base stations, cell sites or radio transmitters, or any concerned equipment and execution of the concerned project by the licensee shall be taken up only after prior approval from local army authorities about specific location of BTS with prior intimation to the licensor and concerned authorities in addition to required clearances.

It is to be noted that the same spectrum band can be allocated for multiple uses provided no interference is occasioned. For example, a large number of devices, including Bluetooth headset, RFID tags and readers, and wireless LANs operate in the band 2400–2483.5 MHz, known as the Industrial, Scientific, and Medical Band. However, the former two use very low power and are operable over a distance of 10 m, whereas the wireless LANs span the entire home. Unless the devices are near each other or transmit/receive at the same time, these can be used over the same frequency range.

Further, some services and the associated devices that operate at low power may be exempt from licensing requirements as they are unlikely to create interference with others operating in the same band. For example, in India, the RF medical devices that operate in 402–405 MHz using a maximum radiated power of 25 micro watt or less (i.e., about 0.00001 of the maximum power of a cellular handset, which is about 2.5 watts) are exempt from the requirement of licensing (DoT 2011)). Similarly any RFID equipment operating in 865–867 MHz (the same range in which some of the mobile cellular systems operate) with a maximum transmitted power of 1 watt is also exempt from licensing.

In addition to recommending the use of various bands of spectrum, the ITU [and other organizations such as Institute of Electrical and Electronics Engineers (IEEE)] provide nomenclature and use of different frequency bands of RF spectrum. An example of the nomenclature is given in Figure 1.5.

Spectrum bands in the region of 2 GHz to 30 GHz are also known as microwave frequencies. Letter designations are also commonly used to denote radio frequencies. Table 1.3 gives the letter-based

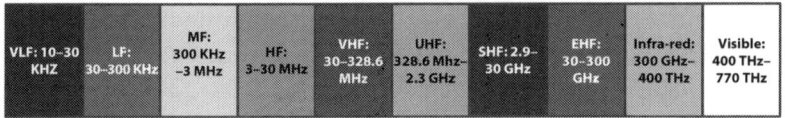

Figure 1.5   Nomenclature of various frequency bands
Source: Authors' own.
Note: VLF: Very Low Frequency; LF: Low Frequency; MF: Medium Frequency;
HF: High Frequency; VHF: Very High Frequency; UHF: Ultra High Frequency;
SHF: Super High Frequency; EHF: Extremely High Frequency. 1 GHz = 10^9 Hz;
1 THz = 10^{12} Hz

Table 1.3   Letter designations of radio frequencies

Letter symbols	Spectrum range in GHz
L	1–2
S	2–4
C	4–8
X	8–12
Ku	12–18
K	18–27
Ka	27–40

Source: Authors' own.

nomenclature. The representation of different bands using band numbers is given in Annexure A1.

Another aspect of the allocation of spectrum bands for specific uses relates to the evolution of telecommunications standards, that is, specifications of a telecommunications system corresponding to a certain set of services and a specific technology. One example of this is the GSM standard in the provision of 2G services evolved for specific spectrum bands such as 900 MHz and 1800 MHz. The allocation of spectrum is also closely related to standards that enable coordination across manufacturers and service providers to trigger economies of scale. It also enables inter-operability of telecom systems. The ITU has a wing referred to as the ITU-Telecom Standardization Sector which is dedicated to the development of the standards referred to. Other standardization bodies such as the European Telecommunications Standards Institute (ETSI) formed in 1988, and Third Generation Partnership Project (3GPP) embrace and evolve certain clusters of spectrum, services, and technologies based on spectrum availability in different countries and technology evolution. Standards organizations also reflect the lobbying efforts of various equipment and device manufacturers.

Early Days of Wireless Mobile Systems

The idea of using radio waves for mobile and personal communications was born in the early 1900s. The early systems were based on the same principles as radio or television broadcasting. They made use of high power transmitters located in base stations on top of the highest point in the coverage area. The transmitters operated at very low frequency levels of around 150 MHz, thus spreading signals over a wide geographical area. A major handicap with these was the low availability of spectrum that was largely allocated to broadcasting and military applications. The few available frequency channels were locked up over a large area and could thus serve only a small number of users. For instance, even as late as 1970, the Bell System in New York could support only 12 simultaneous mobile conversations. Another drawback was the bulky user equipment which made car phones the only feasible mode of mobile communication.

In the early 1920s, both the Marconi Company and Bell Laboratories were testing car-based telephone systems in the US. Bell Labs believes its 1924 system was actually the first two-way, voice-based radio telephone. After the Second World War, the need for improving personal communication became compelling. On 17 June 1946 in Saint Louis, Missouri, AT&T and one of its regional telephone companies, South-western Bell, began operating MTS, or the Mobile Telephone Service. The traffic from the receivers to the transmitter was connected by an operator at a central telephone office. MTS used six channels in the 150 MHz band with 60 kHz-wide channel spacing. Unexpected interference between channels soon forced the Bell System to use only three channels. Waiting lists developed immediately in every one of the 25 cities in which MTS was introduced (Farley 2005).

Outside the US, developments in mobile telephony came slowly. Most governments or Post Telegraph and Telephones (PTTs) did not allow public radiotelephones though there were exceptions. In 1949 the Dutch National radiotelephone network inaugurated the world's first nationwide public radiotelephone system. However, through the 1950s and the 1960s, though pilot tests were conducted using car radio systems and by other methods, the FCC in the US and governments in Europe did not allocate specific radio spectrum for commercial mobile services.

The science behind cell phones as we know them today, was clearly known by 1945 as evidenced by the classic Saturday *Evening Post*, article, 'Phone Me by Air', which quoted Federal Communications Commission (FCC) Commissioner E.K. Jett on frequency re-use for 'small zone systems' that operated at 460 MHz (FCC 2004). The conceptual breakthrough involved creating a network architecture comprising hexagonal cells, each with an immobile transceiver hoisted on a tower. The cellular system made use of low power transmitters, operating at much higher frequency levels than the broadcast systems, typically in the range 400 to 900 MHz. The low power and higher frequencies permitted signals to stay within the radius of the cell. A given spectrum channel in non-adjacent cells could be used simultaneously for carrying different calls. As a user moved from one cell to another, a new channel had to be assigned quickly in order to maintain uninterrupted communication. This required systems that can rapidly process an enormous amount of data. The large-scale production and sharply declining prices of semi-conductors such as microprocessors and memories made the cellular concept feasible by the 1970s. The only remaining barrier was the clearing of frequencies in the 400–900 MHz range. Regulatory reforms for removing previous users (i.e., broadcasters) from this range took several years. The first licences to cellular mobile operators were granted only at the beginning of the 1980s. More details of the cellular architecture are given in Exhibit 1.2.

The early cellular systems transmitted analog signals, that is, they used radio waves that varied in frequency and amplitude. There were a large number of incompatible analog systems existing in parallel in the early days of the cellular industry. These systems used different frequency bands, and had different rates of efficiency.

Exhibit 1.2   The cellular systems and frequency re-use

In 1953, the Bell System's Kenneth Bullington wrote 'Frequency Economy in Mobile Radio Bands' (Bullington 1953). This dull-sounding paper appeared in the *Bell System Technical Journal* circulated around the world. For perhaps the first time in a publicly distributed paper, the 21-page article hinted, although obliquely, at cellular radio principles. Since not enough spectrum was allocated by the regulators for commercial mobile services, as exemplified by the FCC's decisions, Bell Labs worked on optimizing the use of radio spectrum by splitting the geographical areas into many cells, hence the name 'cellular' mobile services. By splitting the given geographical area into many small areas referred to as 'cells', the frequency band in one call could be used in a non-adjacent cell. Non-adjacency eliminates interference, as illustrated in Figure 1.6. By deploying a re-use factor of seven, the assigned radio spectrum band can be divided across seven cells designated as A–G. The inter-cell distance normally varies between 100 m and 5 km. Each cell is equipped with a tower and a Base Transceiver Station (BTS) with a mounted antenna for transmitting and receiving radio signals from the handsets. The handsets communicate with the nearest cellular BTS using suitable algorithms. The BTS is connected to the public telecoms network. When the handset moves from one cell to another, the call is seamlessly 'handed over' to the BTS in the new cell. A separate frequency set is allotted for communication between the handset and the newer BTS. (Today, new technologies like Code Division Multiple Access (CDMA) allow the use of the same frequency even in adjacent cells. This is referred to as a soft handover).

The superior propagation capability of a lower frequency spectrum is more significant in areas where the cell size is large, for example, in rural areas where the inter-cell distance can be as large as 5 km. Such areas are said to be 'coverage-constrained'. In dense urban areas where inter-cell distances are small, the superior propagation capability plays a less important role, although it leads to lower attenuation of the signal when passing through walls. In urban areas, despite making smaller cells to allow greater

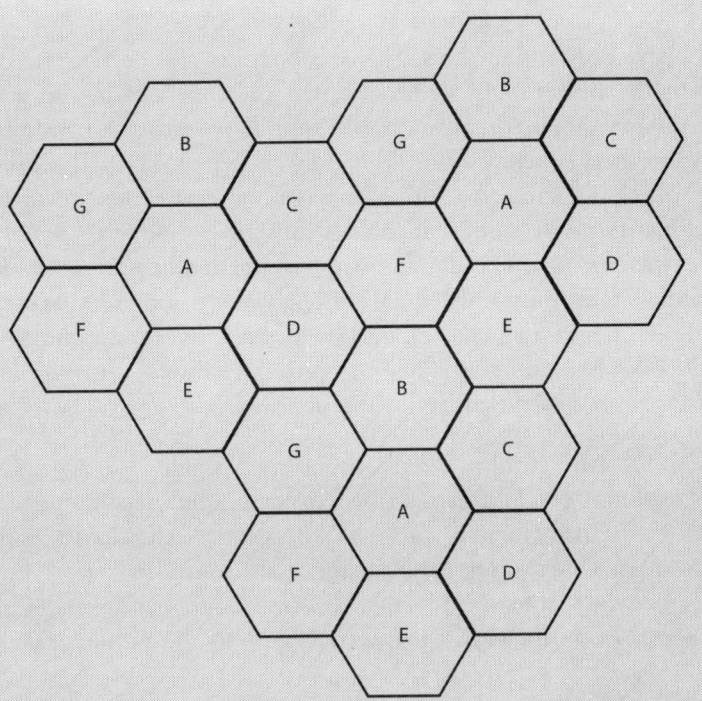

Figure 1.6 A typical cell pattern for re-use in cellular networks
Source: Authors' own.

re-use of frequencies, the problem of spectrum scarcity is often encountered. Such areas are known to be 'capacity-constrained'.

How are the cellular systems and mobile phones different from cordless telephony systems that we use at home? The cordless phones typically operate in the 900 MHz, 1900 MHz, 2400 MHz and, recently, in the 5800 MHz bands (older ones from the 1990s used low frequencies in the range of 34–50 MHz as well). The base station of a cordless phone is in the subscriber's premise, normally connected to a landline telephone outlet and an electrical outlet. The handset communicates to the base station in the designated frequency range. However, the power levels of the base station and handset are set such that the radio links are

(Continued)

Exhibit 1.2   *(Continued)*

available only within the subscriber's premises and do not interfere with the neighbour's cordless system that operates in the same frequency range. Moreover, unlike mobile systems, the handset of the cordless system cannot do handovers across BTSs. This limits it to being a subscriber premise–based system, compared to the cellular mobile system that enables subscriber mobility.

However, of late, there is blurring of the dividing lines between cordless telephony and cellular mobile systems. An example is the origination of Digital Enhanced Cordless Telecommunications (DECT) in Europe as a digital wireless technology in 900 and 1900 MHz to serve the following use cases:

(1) Domestic cordless telephony, using a single base station to connect one or more handsets to the public telecom network.

(2) Enterprise premises cordless PABXs and wireless LANs, using many base stations for coverage. Calls continue as users move between different coverage cells through handovers. Calls can be both within the system and to the public telecom network.

(3) Public access, using large number of base stations to provide a high capacity building or urban area coverage as part of a public telecom network.

Cases 1 and 2 are the most common. Some countries such as India adopted a variation of DECT, referred to as corDECT operating in the 1900 MHz to provide fixed wireless communication service as a substitute for landline connectivity. Apart from reducing the per line cost, corDECT provided both voice and Internet connectivity and was highly appreciated by UNDP as a fast and inexpensive mode of accessing the Internet in developing countries (Sridhar 2012). As per the NFAP [2011; cited in DoT (2011)], the frequency band of 1880–1900 MHz is allotted for the corDECT microcellular system with a maximum transmission power of 250 milliwatts. However, it must be noted that in cases where it is used for fixed wireless communication systems, the base station handovers are not allowed.

This heterogeneity can be explained by the fact that most countries viewed cellular communications as just an additional new business of the state-owned telecommunications monopoly. The development of the cellular network was a means of honing the innovative capabilities of national equipment suppliers.

The most popular systems were Advanced Mobile Phone Service (AMPS) developed in the US, the Nordic Mobile Telephone System (NMT) in the Nordic countries, and the Total Access Communications System (TACS), a British adaptation of AMPS, albeit complying with the different frequency allocation prevailing in Europe. The most successful systems were those that had a large domestic market (AMPS in the US), or those supported by inter-government coordination (NMT for Scandinavian countries). The large diffusion of TACS follows from the fact that it was an adaptation of AMPS to a different band. Examples of C-Netz in West Germany, and NTT in Japan showed that small national markets could not support the development of additional incompatible systems. An episodic account of the development of different national systems is given in Exhibit 1.3.

Exhibit 1.3   Early days of cellular telephony

In January 1969, the Bell System made the commercial cellular radio operational for the first time by employing frequency reuse in a small zone system using public payphones. Passengers on what was called the Metroliner train service running between New York City and Washington DC found they could make telephone calls while moving at more than 160 kmph. Six channels in the 450 MHz band were used again and again in nine zones along the 225 mile route. A computerized control centre in Philadelphia, Pennsylvania, managed the system.

The 1970s and the 1980s saw the rise of Nordic countries in the development of mobile communication systems. Nokia in Finland and Ericsson in Sweden started developing handsets and deploying mobile communication systems in their respective countries.

(Continued)

Exhibit 1.3   *(Continued)*

In May 1978, the Bahrain Telephone Company (Batelco) began operating the first commercial cellular telephone system in the Middle East. The simple two cell scheme had 250 subscribers, operated on 20 channels in the 400 MHz band. In July 1978, AMPS began operating near two American cities. The first area was around AT&T Labs in Newark, New Jersey, and the second was near Chicago, Illinois. Ten cells covering 21,000 square miles made up the Chicago system.

Europe saw the introduction of cellular services in 1981 when the Nordic Mobile Telephone System or NMT450 began operating in Denmark, Sweden, Finland, and Norway in the 450 MHz range. It was the first multinational cellular system. In 1985, Great Britain started using TACS at 900 MHz. Later, the West German C-Netz, the French Radiocom 2000, and the Italian RTMI/RTMS helped make up Europe's nine incompatible analog radio telephone system. In March 1984, the government-owned Korea Mobile Telecommunications (KMT) Company was formed, which began AMPS service in South Korea on 1 May 1984.

2G Mobile Systems

While the world was looking for a breakthrough in mobile technology that could be accepted as a globally harmonized standard, a revolution was brewing in Europe based on the digitization of radio signals, also referred to as Second Generation (2G) radio systems. Digital signals consist of a stream of discontinuous pulses that correspond to the digital bits of zeros and ones used in computers. They can be divided into packets that are transmitted simultaneously with packets from other conversations, to be re-assembled at the point of destination, a phenomenon called multiplexing. This process leads to a significantly more efficient use of spectrum, thereby improving spectrum capacity by a factor of three to six times.

In addition to improving capacity, digital technology protects transmission integrity because digital pulses are more easily regenerated by computers. High transmission integrity allows operators to offer an expanding array of new services including short messaging services (SMS). Finally, digital technology ensures privacy because digital signals cannot be eavesdropped. There are two widely used families of 2G digital systems—Groupe Spécial Mobile (GSM) and CDMA.

Birth of the Ubiquitous GSM

Western Europe had seen a rapid growth in analog mobile telephony service during the 1980s. Most countries had a single system that was not always compatible with the systems of its neighbours. Hence, subscribers had difficulties in roaming. The European Union countries elected to introduce a harmonized standard for 2G or digital service. The belief was that apart from ensuring roaming, a uniform standard would permit greater economies of scale in equipment supply (Gandal et al. 2003).

In 1982, a conference of the Confederation of European Postal and Telecommunications (CEPT) decided that a new digital standard be developed to cope with the increasing demands on European mobile networks. A working party known as the GSM to develop a set of common standards for a pan-European cellular network was formed. The CEPT identified the importance of the availability of common spectrum in the development of a European system and made representations to the European Commission on this issue. This resulted in the EC issuing a directive under which European states were required to set aside spectrum in the 900 MHz band for the future development of a European mobile telecommunications system. In 1987, operators from the CEPT countries signed a memorandum of understanding, usually referred to as the GSM MoU, in which they agreed to deploy the GSM standard at the same frequency in order to facilitate roaming. In 1989, CEPT transferred the GSM committee to the European Telecommunications Standards Institute (ETSI) which completed the specifications of the system.

A rare triumph of European unity, GSM became one of the most convincing demonstrations of what cooperation throughout the European industry could achieve in the global market. Planning began in earnest and continued for several years. In 1990, ETSI published the first recommendations. In 1991, the first GSM call was made in Finland through Radiolinja's mobile network. By this time most of the European operators had adopted GSM as the de-facto mobile communication standard. In 1990, GSM adaptation work for the 1800 MHz spectrum band began. By 1994, there were 1 million GSM subscribers worldwide and included more than 100 operators. The adoption became exponential with GSM subscribers surging past 10 million in the next year.

Australia became the first non-European country to adopt GSM in 1993. In the same year, the first operator network operating in the 1800 MHz band with accompanying smaller handsets was started in the UK. Operators in the US continued to use the digital adaptation of the analog standard—Advanced Mobile Phone System (AMPS). However, in the early 1990s, Personal Communications Service (PCS) was being envisioned as one that describes a set of digital wireless communications capabilities that allows some combination of terminal mobility, personal mobility, and service profile manage-ment. Canada, Mexico, and the US allocated the 1900 MHz band for PCS. In 1995, the first US operator network operating in the 1900 MHz PCS band was inaugurated by the then Vice President of the United States, Al Gore.

As a result of the coordinated adoption of GSM, the technology became mature by the early 1990s, resulting in a stable network with robust features. Its widespread adoption, especially in Europe, led to large network effects for equipment and handset makers and improved roaming compatibility. The Subscriber Identification Module (SIM)–based GSM handsets provided subscribers enough flexibility to switch across networks of choice, leading to the develop-ment of huge prepaid mobile markets in countries such as India. Further, GSM used a relatively open Intellectual Property (IP) rights regime, with firms contributing components to the standards under Fair, Reasonable and Non-Discriminatory (FRAND) conditions. Under FRAND, the firms needed to disclose their IPs and licence it to users at reasonable prices, which in turn reduced the cost of equipment and handsets.

India joined the bandwagon on 31 July 1995, when the first mobile call was made using a GSM phone in the city of Kolkata over Modi Telstra's network. By 1998, thanks to the huge growth witnessed in India and China, the GSM subscriber base hit the 100 million mark. GSM continued to evolve across spectrum bands (for instance, 800 MHz) and technologies (such as General Packet Radio Service and Enhanced Data Gateway Evolution for data communication). Currently, the GSM Association (GSMA), the industry body that supports GSM and its technology evolution path, boasts of 6 billion subscribers.

The CDMA Breakthrough

In July 1985, seven industry veterans came together in the den of Dr Irwin Jacobs' San Diego home in the US to discuss an idea. Those visionaries—Franklin Antonio, Adelia Coffman, Andrew Cohen, Klein Gilhousen, Irwin Jacobs, Andrew Viterbi, and Harvey White—decided that they wanted to build 'Quality Communications' and outlined a plan that has evolved into one of the telecommunications industry's strongest existing technology standards—CDMA.

This complex technology is described very simply by Tanenbaum in his classic book on telecom (Tanenbaum 1996). At the time CDMA was envisioned, the extant 2G technologies used the principle of Time Division Multiple Access (TDMA). This, as Tanenbaum describes, is equivalent to giving each student in a classroom a time slot to speak. The alternative is allowing everyone in the class to speak at the same time, however at a different pitch. This is equivalent to dividing the given frequency band into smaller chunks and giving each to a communication pair in a network, allowing all of them to use their own chunks of spectrum at the same time. This is referred to as Frequency Division Multiple Access (FDMA). The GSM standard uses a combination of FDMA and TDMA.

Irwin Jacobs, who obtained a PhD from the Massachusetts Institute of Technology (MIT) in the US and later taught there, thought about communication differently. He realized it was possible to allow all the students in the class to speak at the same time using the same pitch, however, using different languages (for example, German, English, Hindi, and Portuguese). Using language, it would be possible to differentiate the conversations of the different communicating pairs. CDMA is based on this principle of separating the conversations, not by frequency or time, but by a code. The code, understood both by the sender and receiver, enables both to be able to make sense of the conversation while the conversations of others appear as noise (as German would to a person who knows only Hindi). The advantage of CDMA is that the same set of frequencies can be used in every cell due to its ability to discern signals from noise.

Irwin Jacobs and his co-founders set up Qualcomm in San Diego, US. This company pioneered the CDMA technology and proved its working way back in 1989. However, it was not until the mid-1990s that CDMA really caught on with US operators. The difference in the trajectories of GSM and CDMA was driven by the contrasting approaches of the European and US regulators. While the European regulators mandated GSM as the standard, the US administration followed an approach of freeing up spectrum for digital services and letting market forces decide the choice of standards. There was not even a mandate for nationwide roaming (Gandal et al. 2003).

In 1994, the FCC allocated 60 MHz of new spectrum for digital services. The FCC also allowed the AMPS licence holders to re-farm their spectrum, that is, migrate from analog to digital service. This refarming began in 1995 and much of the AMPS spectrum had been converted by around 2003. By 2003, there was nearly equivalent nationwide coverage in the US for CDMA, DAMPs (the digital successor of AMPS), and GSM with each system being non-interoperable with the other.

Outside the US, GSM was well entrenched as a standard and hence CDMA had a tough time getting accepted as a viable alternative, though many researchers proved that the CDMA's spectral efficiency (that is, the amount of information transmitted per unit of spectrum) was superior to GSM. In 1997, of the 40 million subscribers worldwide, more than 80 per cent were GSM subscribers. Some countries such as South Korea adopted CDMA as the sole standard for 2G network deployment. Others such as India adopted both GSM and CDMA.

The slow uptake of CDMA was also caused by the decision to lock CDMA handsets to specific mobile operators, the IP charges levied by Qualcomm on equipment and handset manufacturers, and roaming incompatibility with non-CDMA networks. However, the inherent advantages of the technology, including higher capacity, lower power usage, and the ability to use the same frequency in adjacent cells resulting in lower call drops caught the attention of standards bodies such as 3GPP. CDMA, after being included under 3GPP (as the 3GPP2 family of standards), became the de facto technical standard for 3G mobile technologies.

Evolution of 2G Standards

The 2G technologies such as GSM and CDMA, primarily designed to provide digital voice services, continued to evolve to support narrow band (that is, low speed) data services as well. Table 1.4 shows the evolution in terms of Kilobites per second (Kbps). The important part of this evolution is that the 2.5G/2.9G technologies used the existing spectrum allocated to an operator for 2G services and hence did not require new allocation of spectrum, thus providing a low cost upgrade path to the operators.

Third Generation Mobile Technologies

Though widely used, GSM has the following disadvantages:

(1) The pulse nature of the TDMA transmission in GSM networks interferes with some electronics including certain audio amplifiers.
(2) GSM has a fixed maximum cell site range of 35 km.

In the early 1990s, the ITU put forth a plan to harmonize the ongoing development of a next-generation wireless network, referred to as International Mobile Telecommunications (IMT) 2000. Under the IMT 2000 initiative, the Third Generation Partnership Project (3GPP) was split into two groups—the Universal Mobile

Table 1.4   Evolution of 2G technologies

GSM family	High-speed Circuit Switched Data (HSCSD)—38.4 Kbps; General Packet Radio Service (GPRS)—144 Kbps; Enhanced Data Rates for GSM Evolution (EDGE)—384 Kbps
CDMA family	IS–95 A and B—115.2 Kbps

Source: Authors' own.

Telecommunication Services (UMTS) group that was formed in 1996 with the sponsorship of ETSI to provide an upgrade to GSM networks and the CDMA 2000 group (3GPP2) that focused on upgrading CDMA networks (Rapport 2002). The UMTS group used the fundamentals of CDMA technology and applied them to wider bands of spectrum (5 MHz as opposed to 1.25 MHz). It also adopted new protocols and specifications and came up with a standard known as W-CDMA (Wideband CDMA). W-CDMA used the same core network as the 2G GSM networks adopted worldwide. It thus gained popularity on the basis on the widespread deployment of GSM networks. The 3GPP2 initiative had to close down as a result. The W-CDMA standard was followed by enhancements including High Speed Packet Access (HSPA), High Speed Downlink Packet Access (HSDPA), High Speed Uplink Packet Access (HSUPA), and HSPA+.

As per IMT 2000, the 3G technologies provide transmission speeds of up to 2.05 Mbps in stationary applications, 384 Kbps per second for slow-moving users, and 128 Kbps for users in vehicles. Thus 3G is considerably faster than 2G and 2.5G technologies. The 3G technologies were designed to support services such as high speed Internet access, video calling, video streaming, and mobile TV.

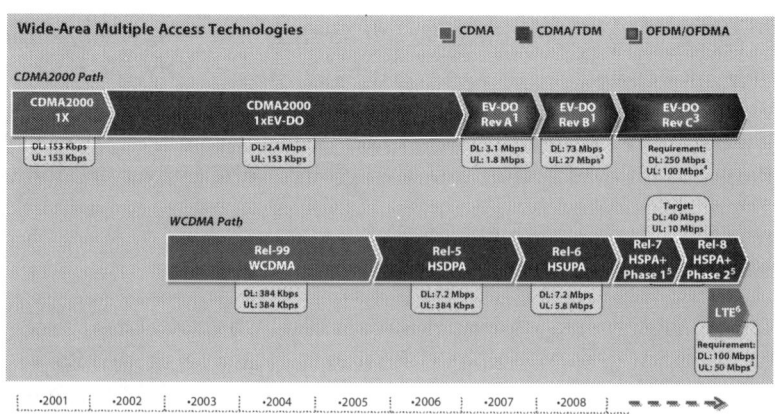

Figure 1.7   Evolution of 3G technologies
Source: Sridhar (2012).
Note: Refer to Glossary for explanation of the terms used here.

Figure 1.7 illustrates the evolution of 3G technologies and the different speeds (Garber 2002).

As on July 2013, there were 514 HSPA networks across 66 countries of which more than 80 per cent supported peak loads of 7.2 Mbps or higher.

Fourth Generation Wireless Networks and beyond

Telecommunications vendors and service providers have started commercial deployment of a next generation, truly broadband wireless cellular system, known as the Fourth Generation System (4G). 4G allows for significantly higher bit rates per user (ranging from 10 Mbps to 100 Mbps), and supports the inter-operability of diverse and heterogeneous wireless and mobile networks. This next generation of wireless technologies promises extensive opportunities for wireless services and applications, namely m-commerce and m-business. The technology is designed to support interactive gaming, high definition (HD) supported video streaming, and mobile TV. The factors that distinguish 4G networks are roaming across networks, interoperability, and higher speeds. In 4G networks, access to multiple wireless networks can also be facilitated by the use of an overlay network or by having intelligence in the networks.

One of the promising standards for 4G was Worldwide Inter-operability for Microwave Access (WiMAX). This technology was standardized by the ongoing work of IEEE802.16 working group under the name Wireless Metropolitan Area Network (WirelessMAN). WiMax was promoted by the WiMAX Forum, an industry association promoting the technology and verifying its compliance with IEEE standards. The core WiMAX standard was developed in 2001 and supported line-of-sight transmission in the 10–66 GHz frequency range. Amendment IEEE 802.16a supporting non–line-of-sight transmission in the range of 2–22 GHz bands was ratified in January 2003. In June 2004, amendment 802.16d—consolidating revisions 'b' and 'c' for quality of service, testing, and inter-operability—was also ratified and is known as IEEE 802.16–2004. Amendment 802.16e–2005, which supports mobility, was concluded in 2005. The first products were certified in January 2006, with average speeds of about 2–4 Mbps. WiMAX functions on both unlicensed and licenced frequencies, but for industrial use the licensed spectrum will be used. The bands used for fixed WiMax include 3.5 GHz and 5.8 GHz and for mobile WiMax include 2.3 GHz, 2.5 GHz, and 3.5 GHz. However, despite its early start, WiMAX did not catch on as it follows a fundamentally different technological approach compared to the extant GSM/WCDMA standards.

Competing with WiMax is Long Term Evolution (LTE) developed by the 3GPP Committee as a natural evolution path for GSM/CDMA networks. Much like GSM and WCDMA, LTE has seen phenomenal adoption by operators. There are two versions of LTE: one that uses paired frequency and splits the given frequency block between uplink (from handset to the base station) and downlink (from base station to the handset) and is referred to as Frequency Division Duplexing (FD)-LTE. The Time Division variant of LTE, referred to as TD-LTE, developed jointly by China Mobile along with the technology firm Qualcomm, is a relatively new entrant. TD-LTE uses a single carrier frequency for both uplink and downlink, dividing the radio frame into sub-frames that can be allocated to either uplink or downlink as per the immediate user need. This dynamic approach is well suited to the changeable data profile of today's users who for the most part download more than they upload, but will occasionally upload data like photos or videos. The use of single frequency instead of splitting the frequency is advantageous to operators who get less spectrum. FD-LTE is being adopted as an evolution path for GSM in the whole of Europe and by most of the US operators, while TD-LTE has fewer takers apart from China. As can be seen in Figure 1.8, technologies such as LTE-Advanced (LTE-A) and IEEE 802.16m offer speeds up to 1 Gbps.

Most often, spectrum is assigned in fragments within a band and even across bands to carriers. Carrier aggregation refers to aggregation of such spectrum fragments to provide an increased channel bandwidth, thus providing increased transmission speeds. This scheme is used in IMT Advanced systems such as LTE-Advanced to provide a transmission speed in excess of 1 Gbps. The aggregated channel can be considered by the terminal as a single enlarged channel from the RF viewpoint.

We are also currently seeing the emergence of what is referred to as '5G technologies' that provide more than 5 Gbps speed and an

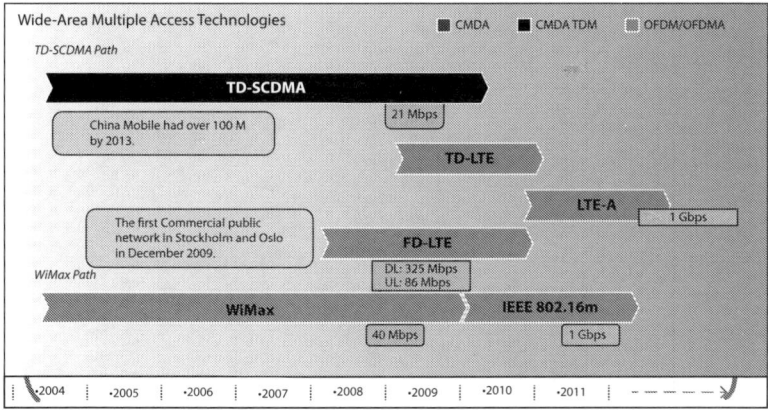

Figure 1.8   Evolution path for 4G
Source: Sridhar (2012).
Note: Refer to Glossary for explanation of the terms used in this figure.

integrated platform for a combination of radio access technologies to coexist. The standardization efforts by 3GPP are expected to start soon and are expected to be completed by 2017. To provide multi-Gigabits speed, these technologies typically operate over a wider channel bandwidth ranging from 100 MHz to even GHz. By widening the channel bandwidth, and using lower power, these technologies necessitate ultra-dense networks with many pico-cells operating at very high speeds, using a combination of licence and unlicensed spectrum.

Table 1.5 provides a comprehensive comparison of 1G through 5G technologies.

Figure 1.9 illustrates in a mobility–data rate graph how these different technologies are positioned.

There is a close relationship between assigned spectrum bands and the technologies that have been implemented in those bands. Though there are world bodies to guide these relationships, individual countries differ in terms of allocation of specific spectrum bands for commercial mobile services due to the constraints that they face. Hence the adoption of technologies in specific bands also varies across countries. However, the wider the adoption of a particular technology in a specific spectrum band, the better are the associated network

Table 1.5 Comparison of various mobile technologies

Technology/Features	1G	2G/2.5G	3G	4G	5G
Start/Deployment	1970/1984	1980/1999	1990/2002	2000/2010	2015/2020
Data Bandwidth	2 Kbps	14.4–64 Kbps	2 Mbps	200 Mbps to 1 Gbps for low mobility	5 Gbps and higher
Standards	AMPS	2G: TDMA, CDMA, GSM 2.5G: GPRS, EDGE, 1*RTT	WCDMA, CDMA-2000	Single unified standard	Single unified standard
Technology	Analog cellular technology	Digital cellular technology	Broad bandwidth CDMA, IP technology	Unified IP and seamless combination of broadband, LAN/WAN/PAN and WLAN	Unified IP and seamless combination of broadband, LAN/WAN/PAN/WLAN and the Internet

Service	Mobile telephony (voice)	2G: Digital voice, short messaging 2.5G: Higher capacity packetized data	Integrated high quality audio, video, and data	Dynamic information access, wearable devices	Dynamic information access, wearable devices
Multiplexing	FDMA	TDMA, CDMA	CDMA	OFDMA	Not yet standarized
Switching	2G: Circuit 2.5G: Circuit for access network and air interface; packet for core network and data	Packet except circuit for air interface	All packet	All packet	
Core Network	PSTN	PSTN	Packet network	Internet	Internet
Handoff	Horizontal	Horizontal	Horizontal	Horizontal and vertical	Horizontal and vertical

Source: Akhtar (2010).
Note: Refer to Glossary for explanation of the terms used in this table.

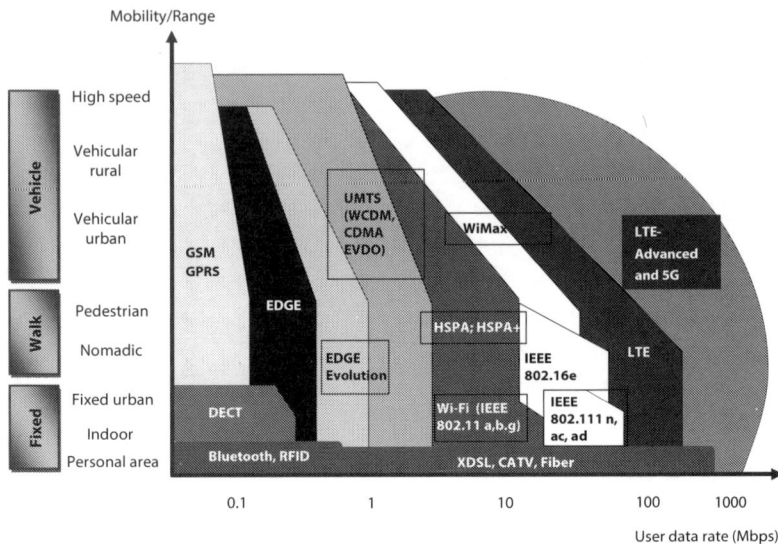

Figure 1.9   Mobility and user data offerings of different technologies
Source: NSN (2007).

effects leading to economies of scale in the associated equipment and handsets.

Table 1.6 illustrates the different frequency bands and the 3G and 4G technologies that have been deployed in those bands in different regions of the world. As can be seen, 2100 MHz band for 3G WCDMA and 2500 MHz for 4G LTE are the most adopted bands across the regions. The CDMA technology deployed in PCS 1900 MHz band (not mentioned in Table 1.6) is unique to the Americas. Deployment of UMTS and LTE in 1500 MHz is unique to Japan. While the spectrum band in 2100 MHz auctioned in 2010 in India has wider adoption for UMTS (that is, the 3G technology of WCDMA), the 2300 MHz for BWA had fewer deployments of 4G technologies such as LTE in that band.

Unique combinations of technology and spectrum are less likely to attain critical mass of adoption, thus losing out on network effects and possibly killing the technology completely. An example is the Time Division-Synchronous Code Division Multiple Access (TD-SCDMA), a variation of CDMA technology promoted by the Chinese government as a possible alternative standard to the 3GPP

Table 1.6 Deployment of 3G and 4G technologies across different spectrum bands in different countries

Spectrum Usage: Device Ecosystem Synergy

Region	Country	700 MHz	800 MHz	900 MHz	1800 MHz	2100 MHz	2300 MHz	2500 MHz
		>2015: LTE-FDD	CDMA-EVDO-Rev B	GSM, >2014: WCDMA*	GSM, LTE-FDD	3G-WCDMA	LTE-TDD	LTE-TDD (Band 41)
R3	India	>2015: LTE-FDD						LTE-TDD (Band 41)
R1	Belgium			GSM				LTE-Band 7/38
	Denmark						>2016	LTE-Band 7/38
	Finland						>2016	LTE-Band 7/38
	France				GSM		>2016	LTE-Band 7/38
	Germany			GSM				LTE-Band 7/38
	Italy						>2016	LTE-Band 7/38
	Norway				GSM		>2016	LTE-Band 7/38
	Spain				GSM		>2016	LTE-Band 7/38
	Sweden				GSM		>2016	LTE-Band 7/38
	UK							LTE-Band 7/38
R2	US	LTE-FDD	CDMA	GSM / Fixed service	GSM			WiMax
	Argentina		CDMA					WiMax
	Russia				GSM/iBurst			
	Brazil	2016	GSM	GSM / ISM	GSM		FDD	
	Canada		GSM					

(Continued)

Table 1.6 (*Continued*)

Spectrum Usage: Device Ecosystem Synergy

Region	Country	700 MHz	800 MHz	900 MHz	1800 MHz	2100 MHz	2300 MHz	2500 MHz
R3	Australia	2014	WCDMA/HSPA+		GSM		802.16	
	China	LTE-TDD	CDMA	GSM	GSM		2015	2015
	Hong Kong	>2014	CDMA/HSPA+					
	Indonesia	>2014	CDMA	GSM	GSM		802.16d	
	Japan	2013	IMT-MC/DS			IMT-MC/DS		Band 41(lower portion)
	Korea	2013					WiBro	
	Malaysia		CDMA		GSM		WiMAX	
	New Zealand	2014	CDMA/WCDMA		GSM		904/910	
	Philippines	>2014	WCDMA/HSPA	GSM			WiMAX	
	Singapore	>2014		GSM	GSM		802.16e	
	Thailand	>2014	CDMA/WCDMA/HSPA					
	Vietnam	>2014	CDMA	GSM	GSM			

Indicates harmonization across countries

Indicates outliers

Source: Authors' own.

approved 3G technologies such as WCDMA and CDMA-EVDO. Though the use of this technology was mandated by the government through the licensing of spectrum in 2300 MHz to China Mobile, the largest mobile operator in China, the adoption of handsets and the associated service was low due to the lack of a supporting ecosystem. Subscribers in China preferred to subscribe to China Unicom's WCDMA service operating in the widely deployed 2100 MHz band, due to roaming compatibility, inter-connection, price, and availability of handsets. Slowly China Mobile started discontinuing TD-SCDMA and invested in building alternate 4G technologies such as TD-LTE.

It can also be seen that the widely used bands of 900 MHz and 1800 MHz for 2G GSM services are being re-deployed for 3G and 4G technologies (also known as spectrum refarming which is discussed in detail in later chapters). As can be seen in Table 1.6, 900 MHz is being used for 3G WCDMA services while 1800 is used for 4G LTE services. Europe also recently harmonized the L-band (1452–1492 MHz) of unpaired spectrum for possible use as supplemental downlink (to offer much greater downlink speed) for LTE technologies.

An interesting question is whether the mandated standards are more beneficial than market determined standards in wireless communications. However, the answer to this question depends on several factors including whether market competition led to technological improvements in wireless technology, whether compatibility (standardization) matters for the adoption of wireless technologies, as well as other regulatory decisions about related factors such as allocation of bands among competing uses, and spectrum refarming mandates (Gandal et al. 2003). This is discussed in Chapter 3.

Wi-Fi and the Spectrum of Commons

As seen in the earlier sections, spectrum for mobile services is always rationed out after much deliberation among government policymakers, regulators, and mobile operators. However, a silent revolution took root in large US university campuses in the 1990s in the form of Wireless Local Area Networks (WLAN) later ratified by manufacturers as Wireless Fidelity (Wi-Fi). Like the Internet and web, wireless LANs became a mass market technology due to open standards that unleashed powerful competitive forces and innovations. While many WLAN technologies were being sold throughout the 1990s, it was the establishment of the 802.11b standard by the IEEE, which set the stage for mass market development. After many other market entries, by the summer of 2002 there were an estimated 15 to 18 million 802.11b networks. Wi-Fi gained popularity as it operated in the Industrial, Scientific and Medical (ISM) band of 2.4–2.4835 GHz (that is, in between licenced S-bands of 40 and 41). There are many devices that operate in the ISM band including cordless phones at home. However, using the ultra wide band of 20 MHz, Wi-Fi uses spread spectrum technology along with low power levels for indoor coverage, thus minimizing interference problems with devices that operate in the same band.

For larger areas or outdoor spaces such as parks and plazas, directional antennas and amplifiers can be used to sculpt a coverage zone using the meagre one watt of power permitted by the FCC for unlicenced operators. As in the case of other open source movements, the proponents of WLAN had the utopian dream of completely undermining the stranglehold of cellular and landline telecommunications companies (Schmidt and Townsend 2003).

No longer having to wait and pay huge sums of money for spectrum, the Internet Service Providers (ISPs), quickly adopted the new technology and started developing Wi-Fi hotspots in public areas, cafes, and restaurants to enable Internet access. The IEEE 802.11 forum with appropriate modifications to the physical layer, encoding algorithms, and multiplexing techniques has kept the evolution of Wi-Fi access networks almost at par with the 2G/3G/4G networks of mobile operators. Table 1.7 illustrates the evolution of Wi-Fi technologies.

Due to changing needs of users (for instance downloading HD videos), there is a need for upgrading the speed of the Wi-Fi networks beyond current levels. The IEEE standard group has been working on the next generation Wi-Fi—IEEE 802.11ac, that is designed to work in the 5 GHz range. The 5 GHz spectrum range has less interference compared to 2.4 GHz in which appliances such as microwave and cordless phones work. Further, using Multi User Multiple Input, Multiple Output (MU-MIMO), the theoretical throughput of IEEE 802.11ac networks has broken the Gigabit per second barrier.

Table 1.7 The evolution of IEEE 802.11 Wi-Fi access network technologies

IEEE 802.11 Standard	Release	Frequency (in GHz)	Channel bandwidth (in MHz)	Max Data Rate (in Mbps)	Indoor Range (in m)	Outdoor Range (in m)
A	Sep 1999	5	20	54	35	120
B	Sep 1999	2.4	20	11	35	140
G	Jun 2003	2.4	20	54	38	140
N	Oct 2009	2.4/5	20/40	72.2/150	70	250
Ac	Nov 2011 (draft)	5	20/40/80/160	87.6/200/433.3/866.7	70	250
ad (WiGig)	Jan 2013	2.4/5/60		7000		

Source: Authors' own.

The standard specifies ultra-bandwidth extending up to 160 MHz thus enabling high data rate capabilities (Qualcomm 2012).

Realizing the benefits of this high-speed Wi-Fi technology, mobile operator forums such as the Third Generation Partnership Project (3GPP) have started adopting it for off-loading part of their cellular network traffic. More details on Wi-Fi offloading are given in later chapters.

On 20 February 2013, the FCC took a major step for relieving congestion caused by huge data traffic in public locations such as airports and hotels. The commission proposed a larger portion of about 135 MHz to be released in the 5 GHz spectrum band for promoting Wi-Fi—IEEE 802.11ac—as a means of providing high bandwidth access (Wyatt 2013). However, some of the proposed band is being used by public and private organizations including the US military. Hence the proposal has already drawn controversies and objections for possible interference. However, the FCC Chairman Julius Genachowski is strongly of the opinion that all the stakeholders will collaborate to solve the interference problems. In later chapters, we reintroduce the topic of spectrum of the commons and address ways of managing spectrum in the commons space.

The newly proposed standard of IEEE 802.11ad, also referred to as WiGig, is expected to provide a speed of several Gbps in tune with the macro cellular 5G technologies.

Future of Wireless Technologies

With the widespread adoption of smartphones and tablets along with bandwidth-hungry applications, it is expected that data will grow 100 to 1,000 times in the future. The networks also need to evolve in keeping with this growth in demand. Apart from the radio frequency–based access link, the backhaul capacity required for transporting the bits from access networks is also being strained.

In this section, we look at future technologies that promise to meet the emerging needs.

Ultra Wide Band and WiGig

In transmitting radio signals, it is possible to trade bandwidth against signal strength. Thus, if more bandwidth is allocated for a signal, it can be transmitted over the same range with a lower signal to noise ratio.[2]

In Ultra Wide Band (UWB), signals can be spread across much of the bandwidth around 3 GHz and 10 GHz. With such a large increase in bandwidth, the power levels can be correspondingly reduced to such a low level that they fall to the limit placed on unwanted emissions from non-communication devices, thus avoiding any interference problems with existing systems (Cave et al. 2011). The FCC has permitted UWB in the frequency range of 3–10 GHz with very low power limits.

Considered as part of the evolution of Wireless Local Area Networks (that is, Wi-Fi) towards 4G, 60 GHz spectrum has been actively considered (similar to UWB) for development. The principal reason for focusing on the 60 GHz band is the huge amount of allotable bandwidth of around 60 GHz, which can be used to accommodate all kinds of short-range (<1 km) wireless communication. Further, the 60 GHz operating frequency has the ability to support these high-rate, unlicenced wireless communications. While unlicensed spectrum in the 2.4 GHz and 5 GHz bands is available internationally, the amount of available 60 GHz bandwidth is an order of magnitude above that available at 2.5 GHz and 5 GHz (Daniels and Heath 2007). However, there are challenges due to directional requirements and high attenuation in the 60 GHz spectrum. Research on modulation schemes, coding/encoding is being carried out to overcome these challenges as this technology is apt for deployment in vehicular networks and ad-hoc short distance networks.

The latest addition to the local area wireless networks specifications is IEEE 802.11ad, also popularly known as WiGig, that uses

---

[2] This stems from the classic paper titled 'The mathematical theory of communication' written by Shannon (1948). Shannon postulated in this paper the following:

$$Max\ Transmission\ Rate\ in\ bps = H log_2(1 + S/N)$$

where, S/N is the signal to noise ratio that determines quality of signal and H is the channel bandwidth. By increasing H, for the same S/N, transmission rate can be increased. Alternatively for a relatively large H, the S/N can be reduced to maintain or even increase the transmission rate. The bandwidth of the signal can be increased by 'spreading' it over wider frequency block.

60 GHz spectrum apart from 2.4 GHz and 5 GHz. The IEEE Standards Board approved the IEEE 802.11ad in January 2013 to provide data rates up to 7 Gbps, more than 10 times the maximum speed previously enabled within the IEEE 802.11 standard. With the improvements introduced in IEEE 802.11ad, this amendment is a perfect complement to the existing IEEE 802.11 standard, acting as the foundation for tri-band networking, wireless docking, wired equivalent data transfer rates, and uncompressed streaming video. The technology allows seamless roaming across 2.4 GHz, 5 GHz, and 60 GHz networks. While it can coexist with Wi-Fi, one competitor to the WiGig Alliance is the WirelessHD (HD—high definition) specification supported by the WirelessHD Consortium, whose multi-Gigabit standard can transfer data at speeds of up to 10 Gbps at a maximum distance of 32 feet.

Pico-Cells and HetNets

As bandwidth hungry applications proliferate, In-Building Solutions (IBS) will augment the capacity of macro cellular networks, thus taking part of the load on spectrum. While the concept of micro-cells and pico-cells has been around for some time now, with the developments of Wi-Fi hotspots and Femtocells (described in later sections), the indoor coverage has expanded. With governments considering more unlicenced bands, especially above 2 GHz, the future networks will have many small cells operating at high frequency spectrum complementing operators' macro cellular networks. Figure 1.10 gives diagrammatic representation of such a small cells architecture.

These IBS not only connect mobile devices but also a variety of other consumer electronic devices such as TVs, music systems, and even home surveillances systems. The Digital Living Network Alliance (DLNA) is an example of this evolution where digital information such as music and video can be shared across devices using the high capacity network (for example, Wi-Fi and Bluetooth) within homes.

Since the small cells can be deployed using licensed/unlicensed spectrum bands and can use any underlying technology (Wi-Fi, Bluetooth, ZigBee, 3G, or 4G), this architecture is generally referred to as HetNets (for Heterogeneous Networks). Realizing the potential of small cells in flexible spectrum and capacity management,

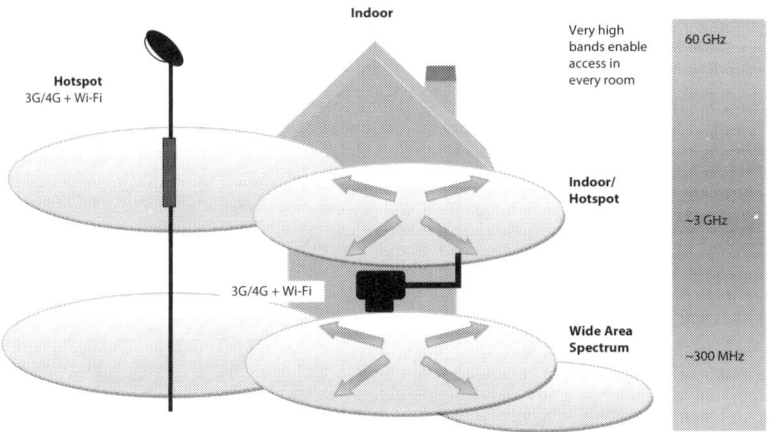

Figure 1.10   Small cells dense architecture
Source: Authors' own.

standard organizations such as 3GPP have incorporated features such as seamless hand-offs between macro cellular networks and smaller IBS networks. Hence we can expect a move from long-range, low-density networks towards short-range, high-density networks with augmented capacity in the near future. This is depicted in Figure 1.11.

The backhaul of these small cells can be Digital Subscriber Link (DSL)–based wireline broadband. Alternatively, in emerging countries such as India where landline penetration is very poor, the backhaul will be through the 4G LTE/WiMax network of a mobile operator. Variations of HetNets are discussed in later chapters.

Dynamic Spectrum Management and Cognitive Radio

Dynamic Spectrum Management (DSM) aims at improving spectrum efficiency by dynamically accessing the spectrum resources based on a real time sensing of vacant channels. The original concept was introduced by Mitola (2000) with the name of Cognitive Radio (CR). He described this concept as a cycle of activities performed by a *radio frequency transceiver*, which included 'observe, orient, plan, learn, decide and act'. Years later, ITU (ITU-R M.2225, 2011) finally defined the main terms related to DSM such as Cognitive Radio System (CRS) and Software-Defined Radio (SDR).

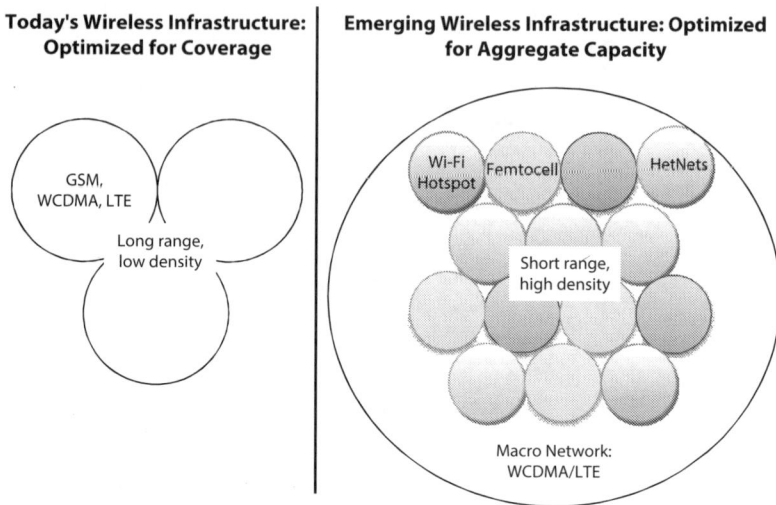

Figure 1.11   Future network architecture
Source: Authors' own.

Despite large efforts in R&D, these technologies have not been successfully introduced in the mobile market. Several reasons have been identified for this slow deployment. The dynamic management of the spectrum involves, in practice, most of the telecommunication layers and not only end user terminals as Mitola originally suggested. In fact, different approaches for using these technologies are seen as possible. Currently, several standards for DSM are under development such as those related to the IEEE and ETSI (for example, IEEE 802.22). In a spectrum-constrained environment, flexible spectrum management using CR is seen as a potential area of research and practice. Details on DSM and CR are discussed in later chapters.

Li-Fi

In the world of scarce radio frequencies, is there any possible hope of getting more spectrum for growing mobile services? One such spectrum band that seems to hold promise is visible light. Researchers have been working to enable computer devices to begin using light to communicate. Though at experimental stages, if it works out, we may one day dump Wi-Fi for something else, perhaps Li-Fi, which is formally called Visible Light Communications (VLC) (Breeden 2012).

Adding a microchip to a standard Light Emitting Diode (LED) light can make it blink millions of times per second. Mobile devices with readers can then translate those blinks, essentially ones and zeros, into data. Adding a LED to mobile devices will allow communication in the other direction.

In this world every street light could become a high-speed Internet port. The human eye would not notice the difference. Much the same way as movies appear to show solid, moving images because the frames are whizzing by at 24 frames per second, nobody will be able to tell the difference between a data-enabled light and a standard, always-on bulb.

Of course, there are some problems. Light communication needs constant line-of-site. Radio waves can travel through a lot of substances without a problem. But if someone walks between a Li-Fi device and a receiver, the communication is broken, not to mention the obvious fact that the signal cannot travel through walls, or anything that can dampen or stop light from passing.

There is also the potential problem of light pollution, especially in cities where this technology will be most useful. Lots of neon signs and non-communicating lights could interfere with the signal. And what happens if two hubs are close together, or a user is walking from one to the next? How will the handoff take place? There is probably a way to go on the idea of Li-Fi replacing standard wireless!

Conclusion

Spectrum, as explained in this chapter, is a valuable public resource essential for today's communication. Technologies for improving capacity and efficiency of wireless systems are developing with great rapidity. However, data hungry applications and smart devices demand ever-increasing bandwidth. This poses interesting problems for technologists, economists, policymakers, regulators and, of course, for end users. In later chapters of this book, we explore the conundrums of spectrum management along the dimensions of technology, economics, and policy and examine the exciting possibilities that the future holds in store.

Annexure A1 Globally harmonized frequency bands for mobile communication services

Operating Band	Uplink (UL) Operating Band Base Station Receive User Equipment Transmit	Downlink (DL) Operating Band Base Station Transmit User Equipment Receive	Duplex Mode	Channel Bandwidths (MHz)	Approximate Centre Frequency
I (1)	1920 to 1980 MHz	2110 to 2170 MHz	FDD	5, 10, 15, 20	2100 MHz
II (2)	1850 to 1910 MHz	1930 to 1990 MHz	FDD	1.4, 3, 5, 10, 15, 20	1900 MHz
III (3)	1710 to 1785 MHz	1805 to 1880 MHz	FDD	1.4, 3, 5, 10, 15, 20	1800 MHz
IV (4)	1710 to 1755 MHz	2110 to 2155 MHz	FDD	1.4, 3, 5, 10, 15, 20	1700 MHz
V (5)	824 to 849 MHz	869 to 894 MHz	FDD	1.4, 3, 5, 10	850 MHz
VI (6)	830 to 840 MHz	875 to 885 MHz	FDD	5, 10	850 MHz
VII (7)	2500 to 2570 MHz	2620 to 2690 MHz	FDD	5, 10, 15, 20	2600 MHz
VIII (8)	880 to 915 MHz	925 to 960 MHz	FDD	1.4, 3, 5, 10	900 MHz
IX (9)	1749.9 to 1784.9 MHz	1844.9 to 1879.9 MHz	FDD	5, 10, 15, 20	1800 MHz
X (10)	1710 to 1770 MHz	2110 to 2170 MHz	FDD	5, 10, 15, 20	1700 MHz
XI (11)	1427.9 to 1447.9 MHz	1475.9 to 1495.9 MHz	FDD	5, 10	1500 MHz
XII (12)	699 to 716 MHz	729 to 746 MHz	FDD	1.4, 3, 5, 10	700 MHz
XIII (13)	777 to 787 MHz	746 to 756 MHz	FDD	5, 10	750 MHz
XIV (14)	788 to 798 MHz	758 to 768 MHz	FDD	5, 10	750 MHz

Band	Uplink	Downlink	Duplex mode	Bandwidths (MHz)	
XVII (17)	704 to 716 MHz	734 to 746 MHz	FDD	5, 10	700 MHz
XVIII (18)	815 to 830 MHz	860 to 875 MHz	FDD	5, 10, 15	850 MHz
XIX (19)	830 to 845 MHz	875 to 890 MHz	FDD	5, 10, 15	850 MHz
XX (20)	832 to 862 MHz	791 to 821 MHz	FDD	5, 10, 15, 20	800 MHz
XXI (21)	1447.9 to 1462.9 MHz	1495.9 to 1510.9 MHz	FDD	5, 10, 15	1500 MHz
XXII (22)	3410 to 3490 MHz	3510 to 3590 MHz	FDD	5, 10, 15, 20	3500 MHz
XXIII (23)	2000 to 2020 MHz	2180 to 2200 MHz	FDD	1.4, 3, 5, 10	2000 MHz
XXIV (24)	1626.5 to 1660.5 MHz	1525 to 1559 MHz	FDD	5, 10	1600 MHz
XXV (25)	1850 to 1915 MHz	1930 to 1995 MHz	FDD	1.4, 3, 5, 10, 15, 20	1900 MHz
XXVI (26)	814 to 849 MHz	859 to 894 MHz	FDD	1.4, 3, 5, 10, 15	850 MHz
XXVII (27)	806 to 824 MHz	851 to 869 MHz	FDD	1.4, 3, 5, 10, 15	850 MHz
XXVIII (28)	703 to 748 MHz	758 to 803 MHz	FDD	5, 10, 15, 20	750 MHz
XXIX (29)	1850 to 1910 MHz or 1710 to 1755 MHz	716 to 728 MHz	FDD	5, 10	700 MHz
unnumbered	2305 to 2315 MHz	2350 to 2360 MHz	FDD	5, 10	2300 MHz
XXXIII (33)	1900 to 1920 MHz		TDD	5, 10, 15, 20	
XXXIV (34)	2010 to 2025 MHz		TDD	5, 10, 15	
XXXV (35)	1850 to 1910 MHz		TDD	1.4, 3, 5, 10, 15, 20	

(Continued)

Annexure A1 *(Continued)*

Operating Band	Uplink (UL) Operating Band Base Station Receive User Equipment Transmit	Downlink (DL) Operating Band Base Station Transmit User Equipment Receive	Duplex Mode	Channel Bandwidths (MHz)	Approximate Centre Frequency
XXXVI (36)	1930 to 1990 MHz		TDD	1.4, 3, 5, 10, 15, 20	
XXXVII (37)	1910 to 1930 MHz		TDD	5, 10, 15, 20	
XXXVIII (38)	2570 to 2620 MHz		TDD	5, 10, 15, 20	
XXXIX (39)	1880 to 1920 MHz		TDD	5, 10, 15, 20	
XL (40)	2300 to 2400 MHz		TDD	5, 10, 15, 20	
XLI (41)	2496 to 2690 MHz		TDD	5, 10, 15, 20	
XLII (42)	3400 to 3600 MHz		TDD	5, 10, 15, 20	
XLIII (43)	3600 to 3800 MHz		TDD	5, 10, 15, 20	
XLIV (44)	703 to 803 MHz		TDD	5, 10, 15, 20	

Source: Huawei (2013).

Note: FDD—Frequency Division Duplexing; TDD—Time Division Duplexing.

References

Aggarwal, V. (2006). 'Jagadish Chandra Bose: The real inventor of Marconi's wireless detector', *Ancient Wireless Association Journal* 47(3): 50–54.

————. (2013). *Sir Jagadish Chandra Bose*. Available at: http://web.mit. edu/varun_ag/www/bose.html. Accessed on 15 March 2013.

Akhtar, S. (2010). '2G-4G Networks: Evolution of technologies, standards, and deployment', *Encyclopaedia of Multimedia Technology and Networking*. Available at: http://faculty.uaeu.ac.ae/s.akhtar/EncyPaper04. pdf. Accessed on 7 April 2013.

Bose, J.C. (1899). 'On a self-recovering coherer and the study of the cohering action of different metals', *Proceeding of The Royal Society* LXV(416): 166–172. (Reprinted, January 1998, *IEEE Proceedings* 86(1): 244–7).

Brain, Marshall. (2000). 'How the radio spectrum works', 1 April. Available at: http://electronics.howstuffworks.com/radio-spectrum.htm. Accessed on 18 December 2012.

Breeden, John. (2012). 'Is the light spectrum next frontier for wireless?' 6 December. Available at: http://gcn.com/blogs/emerging-tech/2012/ 12/light-spectrum-next-wireless-frontier.aspx. Accessed on 11 December 2012.

Bullington, K. (1953). 'Frequency economy in mobile radio bands', *Bell System Technical Journal* 42–62. Available at: http://www.alcatel-lucent. com/bstj/vol32-1953/articles/bstj32-1-42.pdf. Accessed on 7 November 2012.

Cave, M., C. Doyle, and W. Webb. (2011). *Essentials of Modern Spectrum Management*. Cambridge, UK: Cambridge University Press.

Daniels, R., and R. Heath. (2007). '60 GHz wireless communications: Emerging requirements and design recommendations', *IEEE Vehicular Technology Magazine* 2(3) 41–50.

Danigelis, A. (2012). 'Wireless could have saved lives on the *Titanic*', Available at: http://news.discovery.com/tech/titanic-wireless-120411.html. Accessed on 16 June 2012.

Department of Telecommunications (DoT). (2008). 'Amendment to the cellular mobile telephone service licence agreement issued prior to 2001.' Available at: http://www.dot.gov.in/as/cmts_137.pdf. Accessed on 3 January 2013.

————. (2011). National Frequency Allocation Plan (NFAP)-Draft India remarks in the National Frequency Allocation Table. Available at: http://www.dot.gov.in. Accessed on 10 January 2012.

Farley, T. (2005). 'Mobile telephone history', *Telektronikk* 22–34.

FCC. (2004). 'A short history of radio.' Available at: http://transition.fcc.gov/omd/history/radio/documents/short_history.pdf. Accessed on 15 October 2012.

———. (2012). 'FCC online table of frequency allocations.' Available at: http://transition.fcc.gov/oet/spectrum/table/fcctable.pdf. Accessed on 8 December 2012.

Gandal, N., D. Salant, and L. Waverman. (2003). 'Standards in wireless telephone networks', *Telecommunications Policy* 27: 325–32.

Garber, L. (2002). 'Will 3G be the next big wireless technology', *IEEE Computer* 26–32.

Huawei. (2013). 'Whitepaper on Spectrum.' Available at: www.huawei.com/ilink/en/download/HW_204545. Accessed on 15 June 2013.

Jakhu, Ram. (2007). 'International regulatory aspects of radio spectrum management: Implications for developing countries like India' p. 15. Available at: http://www.ictregulationtoolkit.org/Documents/Document/Document/3301. Accessed on 15 May 2013.

Mitola, J. (2000). Cognitive radio: An integrated agent architecture for software defined radio. Doctoral thesis, Royal Institute of Technology (KTH), Sweden. Available at: http://web.it.kth.se/~maguire/jmitola/Mitola_Dissertation8_Integrated.pdf. Accessed on 10 February 2013.

Nokia Siemens Networks (NSN). (2007). *LTE and WiMax Technology and Performance Comparison.* Available at: http://projects.comelec.enst.fr/EW2007/Documents/Comparison_LTE_WiMax_BALL_EW2007.pdf. Accessed on 10 March 2012.

Nuechterlein, J., and P.J. Weiser. (2005). *Digital Crossroads.* Cambridge, MA: MIT Press.

Qualcomm. (2012). 'IEEE 802.11ac: The next generation of Wi-Fi standards' (May). Available at: http://www.qualcomm.com/media/documents/files/ieee802-11ac-the-next-evolution-of-wi-fi.pdf. Accessed on 12 February 2013.

Rapport, T. (2002). *Wireless Communications Principles and Practice.* NJ, USA: Prentice-Hall, Inc.

Schmidt, T., and A. Townsend. (2003). 'Why Wi-Fi wants to be free? *Communications of the ACM* 46(5): 47–52.

Shannon, C.E. (1948). 'A mathematical theory of communication', *Bell System Technical Journal* 27:379–423, 623–56.

Sridhar, V. (2012a). *The Telecom Revolution: Technology, Regulation and Policy.* New Delhi: Oxford University Press India.

Stephenson. M. (2002). 'The Marconi wireless installation in R.M.S. *Titanic*.' Available at: http://marconigraph.com/titanic/wireless/mgy_wireless.html. Accessed on 13 June 2012.

Tanenbaum, A. (1996). *Computer Networks*. NJ, USA: Prentice Hall.

Wyatt, Edward. (2013). 'FCC move to ease wireless congestion', *New York Times*, 20 February.

2

The Economic Classification of Spectrum and the Commons Debate

The classic models have been used to view those who are involved in a Prisoner's Dilemma game or other social dilemmas as always trapped in the situation without capabilities to change the structure themselves. This analytical step was a retrogressive step in the theories used to analyze the human condition. Whether or not the individuals who are in a situation have capacities to transform the external variables affecting their own situation varies dramatically from one situation to the next. It is an empirical condition that varies from situation to situation rather than a logical universality.
—Elinor Ostrom, Nobel Prize Lecture, 8 December 2009

Economists have developed a classification scheme for goods and services and concomitant mechanisms for their efficient allocation. In the context of the debate over the manner in which spectrum ought to be managed—licenced, unlicenced, or a combination thereof—it is necessary to position spectrum within the economic classification of goods.

The Economic Classification of Goods

Goods are classified along two dimensions: excludability and rival-rousness (Musgrave and Musgrave 2011). A good is referred to as excludable if it is 'possible' to exclude users from its use—possible from the viewpoint of social acceptability, economic feasibility, and engineering capability. If it costs an enormous amount of money to create systems to exclude users, or if exclusion is infeasible from a social standpoint, or if the state-of-the-art engineering solutions are unable to create exclusionary systems, then the good is said to be non-excludable. Examples of excludable goods include iron ore, land, and a pair of socks. On the other hand, fishing rights on open seas, or power in rural India, or a pasture which has traditionally been shared by a village community are non-excludable goods.

If the use of a good by a user precludes its consumption by another user, then the good is said to be rivalrous. On the other hand, if the use of a good by one user does not in any way reduce its consumption by another user, then it is said to be non-rivalrous. An ice cream bar or a pen being used at a particular point in time is rivalrous, while a movie in a theatre is non-rivalrous, up to the capacity of the theatre (assuming that an individual's enjoyment of the movie is not diminished by the presence of others). Most goods are not perfectly rivalrous or non-rivalrous but somewhere in between, that is, they are partially rivalrous.

It must be specified that classifying a good along one of the categories often requires a clear specification of the amount of the good in question, the time over which its use is being contemplated, and the number of users in question—in other words, a clear understanding of the context. A tennis court is rivalrous to the extent that two sets of players cannot use it at the same time, but if shared use over time is not a problem, and if simultaneous use can be eschewed, then it is amenable to being used by multiple users without reducing anyone's enjoyment. In fact, having more users can often increase each one's enjoyment up to a point, as the number of potential partners increases.

Further, the nature of a good also changes with changing technology. For instance, with the development of low-cost remote surveillance devices, land in far-flung locations can be considered

Table 2.1   The economic classification of goods

	Rivalrous	Non-rivalrous
Excludable	Private goods, for example, ice cream	Toll goods, for example, roads
Non-excludable	Commons goods, for example, village pastures	Public goods, for example, national defence

Source: Authors' own.

an excludable good, although without such technology, it would be considered vulnerable to squatting and, therefore, non-excludable.

Combining the two independent dimensions of excludability and rivalrousness, all goods in specific contexts can be classified in one of the following categories (see Table 2.1):

Excludable and rivalrous: private goods. Examples include items of clothing and food—anything for which private property rights can be defined and which cannot be shared without reducing each individual's enjoyment. Such goods can be efficiently allocated with the use of market mechanisms, assuming certain necessary conditions for competition hold. Of course, while market mechanisms are adequate for achieving efficiency, they may not achieve social norms on equity.

Non-excludable and rivalrous: commons goods, with the name adopting the metaphor of common grazing areas. These are goods where exclusion is not feasible and a user's consumption partly curtails the possibility of consumption by others. Examples include village pastures, ponds, and fishing rights in open seas. The market mechanism breaks down in the case of such goods as potential users would try to 'free-ride' on account of the non-excludability. Unregulated use of commons goods also leads to overuse, a phenomenon referred to as the tragedy of the commons. Such goods were believed to require privatization or government management. However the work of Nobel Prize winner Ostrom (1990) has shown that it is possible for appropriate community institutions to lead to an efficient allocation of commons goods. This finding has great relevance for the discourse on spectrum management.

Excludable and non-rivalrous: toll goods, with the name adopting the metaphor of toll roads. These are goods for which exclusion is possible and users can undertake consumption without curtailing the consumption of others (up to a point). Examples include toll roads, social clubs, and Internet access at times when traffic is low enough to allow additional users without reducing access speeds. The market mechanism breaks down for such goods because, given the non-rivalrousness, it is inappropriate to charge users (even though it is feasible on account of excludability), and hence not possible to recover costs. Such goods are often provided through a model of public-private partnership, where a private party is contracted by the government through competitive bidding and compensated through general budgetary provision. When the toll good starts getting congested, a toll may be levied to bring about efficient usage. Regulatory supervision is required to ensure that the returns to the private entrepreneur are not excessive.

Non-excludable and non-rivalrous: pure public goods. The classic example of a public good is a lighthouse that uniformly gives light to all, and cannot be limited to some seafarers alone. Other examples include national defence and basic research. Such goods are subject to the free-rider problem associated with non-excludability as well as the inappropriateness of levying fees associated with non-rivalrous goods. The large literature on the efficient provision of public goods lists government provision, and provision of subsidies to private producers through competitive bidding as some of the mechanisms to efficiently allocate public goods.

The Economic Classification of Spectrum

We now analyse the nature of spectrum in terms of the four-fold classification.

Excludability: Technological solutions for monitoring exclusive use of spectrum blocks are available. Licensing the use of spectrum is also commonly practiced, having acquired acceptability following the disaster of the *Titanic* (Carter et al. 2003). Therefore, spectrum can be considered an excludable good.

Rivalrousness: The rivalrousness of spectrum is contingent on a number of factors. Users of telephony (or data services) use specific frequencies to transmit signals. The transmission of signals of sufficiently high power by multiple users at the same time and at the same frequency in close physical proximity leads to interference that can render the systems unusable. In some cases, signals may interfere with each other even if they are hundreds of miles apart. Even if users transmit on neighbouring frequencies, they can still interfere. With real life transmitters, the signals transmitted on one channel leak into adjacent channels and with real life receivers, signals in adjacent channels cannot be completely removed from the wanted signal. This interference makes spectrum use rivalrous, at least partially.

However, the degree of interference depends on technology, device capability, number of users, and application needs. The geographical and temporal pattern of peaks and lows in traffic is also significant.

Time Division Multiple Access (TDMA) systems send powerful signals over narrow spectrum channels. In such cases, interference is a real possibility. Multiple users attempting to send signals at the same time, at the same location, with a sufficiently high power would degrade each other's signal. However, today devices use sophisticated multiplexing arrangements that significantly reduce the risk of interference with other devices, even high power ones, using the same frequency bands. These include spread spectrum and ultra wideband techniques where the initial narrow band signal is spread over a wide band of spectrum resulting in a low power level at any given frequency. In frequency hopping spread spectrum, the transmitted signal hops from one frequency channel to another. The receiver hops in strict conjunction with the transmitter, thereby collecting all the data transmitted. The amount of time that the signal is present in any channel is usually very short, commonly less than 10 milliseconds. This reduces the effects of interference both to and from conventional users.

The degree of interference also depends on the quality of the receivers. The development of advanced devices that have higher capability to filter out unwanted signals mitigates the problem of interference.

The development of new applications that need higher levels of speed in transmission and are less tolerant of degradation, for example videoconferencing, increases the relevance and threat of interference.

The regional pattern of traffic enables the avoidance of interference. It has been seen that demand for spectrum is usually localized in central business districts and residential apartments. In order to handle this demand, local nodes are used to provide capacity. These nodes require low power as opposed to towers which transmit over a wide area from a central location. This makes for a lower risk of interference.

Several studies have also shown that most bands are not heavily used at all times. Hence the dimension of time can be used, in addition to frequency, power, and space, in order to permit dynamic allocation and assignment of spectrum usage rights. Time-sharing of spectrum between multiple users would lead to more efficient use of the spectrum resource and reduce the rivalrousness of spectrum use.

While technology plays an important role in determining the economic nature of spectrum and therefore its management framework, the pace of technological change may in turn be affected by the prevailing regime of spectrum management. For instance, allowing spectrum to be treated as a common property resource could well incentivize the creation of technologies that reduce the rivalrousness of spectrum. Necessity may indeed have been the mother of invention in the case of the emergence of new technologies in the context of Wi-Fi. As outlined later, there is an emerging school of thought that advocates treating spectrum as a common property resource, rather than allocating private property rights for its use, partly on the basis of the innovations stemming from the use of spectrum bands treated as a common property resource. The scenarios that could emerge when a partially rivalrous good is treated as a common property resource are now presented.

The Tragedy of the Commons

A partially rivalrous good treated as a common property resource is subject to the tragedy of the commons, or *congestion*. This possibility can endure in the presence of technological progress. A simple example can be provided in the form of a common pasture in a village where five people can potentially graze their cattle. Each head of cattle costs Rs. 100, which can be raised at a 13 per cent rate of interest. Thus Rs. 13 is the marginal cost of putting another cattle to pasture.

The value of the cattle at the end of a year depends on the number of cattle put out to graze. As the number increases, the value reduces

Table 2.2   Overcrowding on a common pasture

No. of cattle	Price per cattle at the end of 1 year	Average revenue per cattle	Total revenue from grazing	Marginal revenue (increase in total revenue for additional cattle)	Excess of marginal revenue over interest due to bank
1	126	26	26	26	13
2	119	19	38	12	-1
3	114	14	42	4	-9
4	111	11	44	2	-11
5	109	9	45	1	-12

Source: Authors' own.

due to the lower availability of pasture per cattle. This reflects the assumption of diminishing returns or partial rivalrousness. The relation between the number of cattle and the value at the end of a year is given in Table 2.2.

The overall profit is maximized when marginal revenue is greater than or equal to the marginal cost. If grazing occurs beyond this point, that is, at levels where the marginal cost becomes greater than the marginal revenue, then the user's surplus is being reduced. In the example, surplus is maximized when one cattle grazes, since the addition of the second cattle makes the marginal revenue lower than the interest cost of Rs. 13.

If individuals make decisions in an uncoordinated manner with the aim of maximizing individual returns, then three cattle will graze on the pasture, since each will make a profit of Rs. 14 which is greater than their interest cost. This phenomenon of over-utilization of a partially rivalrous good in the absence of coordination is referred to as the tragedy of the commons and occurs because users graze till *average revenue* is greater than marginal cost.

Resolving the Tragedy of the Commons

One solution to the tragedy of the commons comes from the conclusions of the Coase Theorem (Coase 1960) which states that if the parties involved can costlessly communicate with each other, then they can arrive at an efficient solution. In the example given earlier, one herdsman will be able to convince the others to give him exclusive rights to the pasture in return for some kind of payment. However, communication costs can be high, and gains may not be transferable. Hence, Coase's sanguine conclusions may not apply to real world scenarios.

Two longstanding solutions are government control and privatization. In the former, the government controls the good and permits grazing till the optimal point. In the case of the latter, the private owner does so. In our example, the pasture would be sold to one person. Since the single owner is earning Rs. 26—an additional Rs. 13 over the bank rate of interest—the maximum price they would be willing to pay in a competitive market would be Rs. 100. The interest cost on Rs. 100 would wipe out the additional return generated through exclusive use of the pasture.

If the pasture is under government control, the government has to take a decision regarding the optimal amount of grazing. Even in order to give the pasture to private control, a prior decision on the optimal number of plots, the pasture has to be divided into has to be made.

In cases where there are a large number of users of the toll good, definition of private property rights may be cumbersome. For instance, a road being used by a 100,000 vehicles daily may not permit such allocation of private property rights. Instead, the toll good may be given to a manager who levies a toll on users in order to prevent overcrowding. The purpose of the toll is to increase the cost so that users make the same decision as they would have had they taken the marginal revenue into account rather than the average revenue. In the example, a toll slightly greater than Rs. 6 would ensure that the second person does not take his cattle to pasture, since his cost (including the interest of Rs. 13) becomes greater than Rs. 19, which is his revenue when two cattle graze. The maximum toll that could be charged would be slightly less than Rs. 13. At a toll above Rs. 13, even the first herdsman would find grazing unviable since the total cost becomes greater than the revenue with one cattle on the pasture. Hence, a toll strictly in between Rs. 6 and Rs. 13 is consistent with achieving an efficient solution for one herdsman. This is called the range of the efficient toll. A low toll falling in this range will yield profits to the users. Competition would raise the toll to the upper bound of the efficient toll range.

In addition to alleviating congestion, the toll also serves to defray the cost of setting up and maintaining the toll good. If the cost is greater than the toll, then the toll good becomes unviable. Nevertheless, if public policy considerations warrant the provision of the good, provisions in the general budget may need to be created. If the cost is lower than the toll, then the levies can be used for R&D to enhance the capacity of the toll good.

If the users of a common property resource are heterogeneous, then competition within each category of users could lead to a heterogeneous structure of tolls that reflects the differing profitability accruing to different classes of users.

The case of spectrum as a toll good presents an additional complexity as, due to the lack of dynamic sensing of signals, there could be congestion even in the absence of rivalrousness. Given the lack of such a 'line of sight' in the case of wireless technologies and the rivalrousness of spectrum, the allocation of private property rights over spectrum blocks has been considered necessary for efficient management. This practice began in the US in the 1930s following the proliferation of devices and the increasing incidence of interference (Carter et al. 2003). Ironically, it was Coase (1959) who, one year before the publication of his celebrated paper enunciating the Coase Theorem, suggested the use of auctions as a way of allocating property rights in spectrum. The presence of new technologies that dynamically choose vacant frequency channels ameliorates this special challenge of spectrum management to some degree.

Capacity-enhancing Technological Progress

Some people claim that new technologies have considerably reduced the risk of interference. Using the example developed earlier in this chapter, we examine the impact of technological advances that increase the capacity of a toll good on the phenomenon of overcrowding in the commons.

In case the capacity increases, the optimal number of users of the toll good may increase. However, unless the increase in the capacity of the new technology is sufficiently large, the prospect of overcrowding

Table 2.3   Overcrowding on a common pasture despite enhanced capacity

No. of cattle	Price per cattle at the end of 1 year	Average revenue per cattle	Total revenue from grazing	Marginal revenue (increase in total revenue for additional cattle)	Excess of marginal revenue over interest due to bank
1	132	32	32	32	19
2	127	27	54	22	9
3	117	17	51	-3	-16
4	112	12 ·	48	-3	-16
5	109	9	45	-3	-16

Source: Authors' own.

remains and the levy of a toll will continue to be necessary. The size of the toll may be lower than, equal to, or greater than the toll with the lower capacity depending on the specific values of the relevant parameters. Table 2.3 shows a possible scenario with enhanced capacity.

The revenues at the end of year 1 have increased reflecting technological advances. If people act in an uncoordinated manner, there would be three cattle put to pasture. However, social surplus is maximized with only two cattle. A toll greater than Rs. 4 and less than or equal to Rs. 14 would ensure that the socially optimal outcome is observed. With competition, a toll close to Rs. 14 would be observed. Note that this is higher than the toll with lesser capacity. In general, it is possible (although not necessary) that the toll goes up with higher productivity, albeit with higher levels of utilization of the good.

The overall social surplus with higher productivity is higher— Rs. 28 instead of Rs. 13 (since each of the grazers makes a profit of Rs. 14 over the interest cost of Rs. 13, while earlier the sole grazer had a profit of Rs. 13 over the interest cost; those who don't graze break even and don't make a surplus).

However, moving from the old scenario to a new one may require prior investments in R&D. Recall that the excess of the toll over the cost of build-out and maintenance was being used to carry out R&D. The amount of funds available in the scenario with the lower capacity may not be sufficient to fund the R&D required to move to the new scenario with enhanced capacity. The extra cost of R&D must be subtracted from the surplus accruing in the new situation. After factoring the extra R&D costs, the total surplus could even go down. If the new technology has already been developed, this will not be a consideration. But sunk costs incurred with the old technology, which may have to be foregone in the migration to the new technology, have to be factored.

Even more vexing questions relate to the manner in which technology will be chosen. Capacity increase can come from discontinuous shifts in technology that render old technologies obsolete and throw up a new ecosystem with its own set of winners and losers. For instance, moving to an unlicensed spectrum will advantage device manufacturers at the cost of those who have built mobile networks using licensed spectrum. The new technologies may also be linked to specific uses which may be different from the current patterns of utilization. Bureaucratic intervention in matters of technology selection is generally frowned upon and this presents a challenge to those who believe that a spectrum commons should be mandated by public policy.

Finally, while in a regime of privatization, toll collections or auction proceeds can be used to fund R&D expenditures, in a commons there is no direct provision for generating a fund which can be used for R&D. This may present a challenge for continued technological progress. On the other hand, the congestion experienced by users in a commons acts as a spur for further innovation, unlike in the case where a toll is levied to eliminate overcrowding.

In brief, the debate on the spectrum commons must recognize realities related to the continuing threat of congestion and the concomitant need for limiting access, the cost of R&D, the manner in which technology is selected, and the manner of choosing the use to which the common property resource is to be put. The debate must also consider the ability of the two different regimes to spur innovation.

Community Solutions

Assigning property rights in spectrum requires a range of administrative decisions including the number of blocks, the tenure of the licence, the power limits of devices, and the geographical area of transmission in each band. The need for the government to take such decisions opens up the possibility that the tragedy of the commons may be supplanted by the spectre of administrative failure.

A new conceptual direction to the management of partially rivalrous resources comes from the work of Ostrom (1990), winner of the Nobel Prize in Economics for her empirical work on common property resources. Such resources are non-excludable but rivalrous. Ostrom's work showing how the problem of rivalrousness in the presence of non-excludability can be addressed without government ownership or private property rights gives pointers on how spectrum, a rivalrous and excludable resource, can be similarly managed.

Ostrom found that certain communities jointly owning the resource were able to create institutional arrangements that influenced and modified the self-serving behaviour of individuals resulting in efficient outcomes. She famously observed that the Mongolian steppes that were neither under private ownership nor under government control continued to remain green belying the prediction of overgrazing by self-interested parties in the presence of rivalrousness and non-excludability. This led her to study the set of institutional principles that resulted in optimal utilization of common property resources by community-based institutions.

The design principles identified by Ostrom are:

- Clearly defined boundaries (effective exclusion of external unentitled parties);
- Rules regarding the appropriation and provision of common resources adapted to local conditions;
- Collective-choice arrangements allow most resource appropriators to participate in the decision-making process;
- Effective monitoring by monitors who are part of or accountable to the appropriators;
- Graduated sanctions for resource appropriators who violate community rules;
- Cheap and easily accessed mechanisms of conflict resolution;
- Recognition of the self-determination of the community by higher-level authorities; and

- In the case of larger common-pool resources: organization in the form of multiple layers of nested enterprises, with small local CPRs at the base level.

Ostrom found that accountability in community-managed systems is stronger than accountability in bureaucratically-managed systems. Thus, from the literature we can conclude that even when a good is partially rivalrous, the possibility of its being efficiently used by a community cannot be ruled out.

The Debate over the Spectrum Commons

Having presented the conceptual aspects of the management of toll goods, we turn to specific issues related to the spectrum commons.

The ideas of 'open access' and 'commons' are often conflated in the licenced–unlicenced debate, and the two words are used interchangeably. But they are two separate concepts with distinct meanings.

Open access is a regime under which everyone has access to an unowned resource without limitation; no one controls access to the resource under open access. Access to sunshine, for example, is open. A commons, on the other hand, is a resource that is managed jointly by a group of individuals. This is sometimes referred to as the 'managed commons'.

The rationale for open access arises from the Coasian view that negotiations can lead to efficient allocation of a common property resource obviating the need for a governance structure. The rationale for a managed commons free of government intervention or the market mechanism stems from Ostrom's view that communities can evolve institutions to self-govern common property resources.

As mentioned earlier, in the case of spectrum, the large number of widely divergent users makes the transaction costs of negotiation significant. Hence, it is the possibility of a managed commons that is relevant, not that of open access. There are two variations of the managed spectrum commons: the first in which no licences are granted but certain rules of the game are specified including power limits. The second in which non-exclusive licences are granted to a limited number of users based on the nature of their services and transmission needs.

Further the spectrum being used in a non-exclusive manner may either be unlicenced (for example, the 2.4 GHz band) or licenced but treated as a 'private commons'. In the managed commons, users can be subject to interference from other users.

Licensed Spectrum

The privatization of a rivalrous resource enables the problem of interference to be internalized. The owner has the incentive to make investments without fear that his investments will be appropriated by others. In the case of spectrum, the possession of a licence ensures that only the owner can use the frequency band, thus limiting interference and ensuring that the owner has the incentive to make complementary investments in building out a network. Through the sale of the licence, the government is able to share the revenue that is generated.

However, the system of allocation of spectrum licences comes with several drawbacks, one of the most important of which is that private spectrum holding can stifle third-party innovation as third-party innovators face the threat of a hold-up (Milgrom et al. 2011). A company that comes up with a new mobile device or business model needs to convince the owner of the spectrum to let it develop its idea and it may have to share a large fraction of the value that is created with the spectrum owner. Permission is unlikely to be given if the new development threatens the owner's existing business. And, if the innovation requires the assent and coordination of multiple spectrum owners, it is even more difficult to get all the owners to agree. The potential for this type of coordination failure is sometimes called the *tragedy of the anti-commons*. It has become a familiar problem in intellectual property, arising when developers must licence such a large number of patents that the process of innovation becomes cumbersome, unwieldy, and inefficient. Overall, for radio spectrum, history suggests a mixed innovation and investment story with licensed spectrum being valuable in encouraging the necessary network infrastructure for wireless mobile handsets and unlicensed spectrum encouraging a long series of novel, valuable, and unanticipated uses. The increased pace of innovation followed by unlicensed spectrum is demonstrated by the early appearance of technologies in

WLAN relative to their introduction in cellular networks. Such technologies include digital signal coding, spread spectrum techniques, and orthogonal frequency division multiplexing. The adoption of an unlicensed spectrum appears to have fostered the development of technologies that facilitate non-rivalrous use of spectrum.

Another concern with allocating property rights is that once licences have been allocated, reallocation tends to be a slow and difficult process. While theoretically spectrum trading can allow reallocation, experience shows that despite the promise of trading, secondary spectrum markets remain thin. In the United States, the creation of nationwide networks from the initially fragmented licence allocation took many years and major corporate acquisitions and consolidations to accomplish.

Finally, increasingly flexible spectrum policies enable spectrum licence holders to seek more valuable uses for existing spectrum. Consequently, they may seek to deploy systems that may be incompatible with the adjoining spectrum users. Regulatory flexibility will provide more access to spectrum, but the inevitable consequence will be conflicts about these more flexible uses.

The Spectrum Commons

The availability of unlicensed spectrum can make the cost of setting up and deploying systems for local wireless transmission extremely low. A lack of entry barriers encourages companies to develop new products and business models and market them directly to consumers rather than having to work with the limited number of licensees that control cellular networks. This has facilitated the rapid introduction of Wi-Fi capabilities to new products: laptops, book readers, tablet computers, and smartphones. The need to solve the problem of interference without recourse to property rights has led to the development of new technological models. Today, services that operate on unlicensed spectrum increasingly compete with services relying on licensed spectrum. For example, voice calls on Wi-Fi networks—in applications like Skype—compete with traditional cellular networks.

More unlicensed spectrum need not mean lower revenues for the government. To the extent that auction payouts by telecom operators are directly linked with the budgets they allocate in advance, changes in available spectrum may not have a significant impact on revenue. Further, if the demand for licensed spectrum is inelastic (that is, less than 1), then a reduction in the spectrum available will increase the revenue from auctions.

In contrast to those who say that the government should provide more unlicensed spectrum are those who believe that licensed spectrum along with flexible use is the way to go, especially in congested urban areas.

The application of the Ostrom principles, that enable the efficient use of a rivalrous resource without private property rights or government intervention in the context of spectrum sharing is subject to a number of qualifications. In the case of a traditional common property resource like a pasture, the identification of the members of a community can be based on tenure of stay in the region. However, the potential users of a spectrum commons include all the devices that could operate in the said frequencies, now and in the future. Due to the large number of such users, it becomes difficult to identify the members of the community who can use the spectrum commons. Further, the community is dynamic as new users can emerge who need to become part of the community on account of the quality and usefulness of their innovations. The government usually needs to intervene to define standards including power limits that devices need to satisfy and thereby limit the potential members of the community.

The proponents of property rights with flexible usage believe that the role of administrative processes in setting norms for the use of unlicensed spectrum, including the specification of power limits, significantly influences the kinds of devices and actors that would emerge. Thus, the drawbacks associated with government intervention continue to be inherited in a regime of unlicensed spectrum. In a partially liberalized regime, the government could be culpable for errors of judgment on the allocation of spectrum to specific uses, or the specification of certain technologies. In the case of unlicensed spectrum, the specification of various parameters, including power limits, device types, and administrative protocols, can amount to a similar misjudgment. The creation of a spectrum commons is associated in the minds of some with the prospect of a return to command and control spectrum management paradigms or a reversion to the days of 'industrial policy'. The tension between those who want 'spectrum for all' albeit with a greater role for the government and those who want flexible spectrum use with minimal restrictions, albeit with a greater risk of conflict between licensees is the central dilemma of spectrum management.

With unlicensed spectrum the onus of value creation shifts to device manufacturers who develop devices to optimize the use of unlicenced bands. However, the difficulties in providing high bandwidth and low latency services using unlicensed spectrum reduces the incentives for innovation in these areas in the context of unlicenced bands. In the context of the 3650 MHz band in the US, some device manufacturers like Intel argue that because a high quality of service cannot be guaranteed over the commons, investments for providing broadband services in the band will be severely constrained. Even if a certain quality of service can be achieved, since the potential number of users of the commons is unlimited, there is no assurance that it can be maintained in the future. This lack of business certainty, manufacturers argue, will prevent significant investments in the band.

A Middle View

There are three possible approaches to spectrum policy: (a) An exclusive rights or flexible use approach: a licensing model in which a licensee has exclusive and transferable use rights for a specified spectrum within a defined geographical area, with flexible use rights that are governed primarily by technical rules to protect users against interference. (b) Commons model: allows an unlimited number of unlicenced users to share frequencies with usage rights that are governed by technical standards but with no right to protection from interference. (c) Command and control model: the traditional process of spectrum management in which allowable spectrum uses by licenced users are limited based on administrative judgments.

A report to US President Barack Obama made by the President's Council of Advisors on Science and Technology (2012) suggests that while command and control dominates today's practice, future spectrum management will be based on a combination of an exclusive use approach, a commons approach, and to a limited degree a command and control approach (in instances where there are compelling public policy reasons such as some public safety application). Modern spectrum technology also permits the allocation of tiered usage rights with primary, secondary, and tertiary users. Aspects of such management regimes are covered in Chapter 7. Such tiered sharing may become an integral part of spectrum usage in the future.

In urban areas, a low-frequency spectrum can be predominantly assigned for exclusive use, while leaving a narrow band of spectrum for the commons to be used to provide municipal Wi-Fi and capacity enhancements in high demand zones like office buildings. Higher frequency bands can be allotted for a commons approach. In rural areas, most spectrum, in both low- and high-frequency bands, can be allocated for the spectrum commons. A small proportion of low-frequency spectrum can be allocated for exclusive use.

Both licensed spectrum and the spectrum commons play a role in today's scenario. Given the scarcity in the capacity of licensed spectrum networks in the context of galloping demand for data services, Wi-Fi is also playing a complementary role to licenced networks. For instance, an iPad owner uses Wi-Fi when it is available, but switches to 3G mobile to operate outside the Wi-Fi hotspots. It is estimated that about a third of all mobile data traffic is routed through Wi-Fi. The Wi-Fi capability creates the basic demand for iPad services for these users with additional mobile data services provided using licensed spectrum. Thus, the availability of unlicensed spectrum applications creates more consumer demand for licensed spectrum services.

Conclusion

The nature of spectrum as an economic good opens up a series of questions that challenge our traditional view of market-oriented societies based on private property. The feasibility of non-exclusive use, the associated challenges, and the immense benefits that can be gained by society upon a wise resolution of the conundrums, make spectrum management an immensely important task. Learning from the management of spectrum can spread to other domains of economic activity and thus, affect the nature of our economic interactions. The future is wide open.

References

Carter, Kenneth, Ahmed Lahjouji, and Neal McNeil. (2003). 'Unlicenced and unshackled: A joint OSP-OET white paper on unlicenced devices and their regulatory issues.' Available at http://www.fcc.gov/working-papers/unlicenced-and-unshackled-joint-osp-oet-white-paper-unlicenced-devices-and-their-regu. Accessed on 10 August 2013.

Coase, Ronald H. (1959). 'The federal communications commission', *Journal of Law and Economics* 2(October): 1–40.

———. (October 1960). 'The problem of social cost', *Journal of Law & Economics* III: 1–44.

Milgrom, Paul, Jonathan Levin, and Assaf Eilat. (2011). 'The case for unlicenced spectrum.' Available at: *SSRN 1948257.*

Musgrave, Richard A., and Peggy B. Musgrave. (2011). *Public Finance in Theory and Practice.* Fifth Edition, New Delhi, India: Tata McGraw-Hill.

Ostrom, Elinor. (1990). *Governing the Commons: The Evolution of Institutions for Collective Action.* UK and USA: Cambridge University Press.

President's Council of Advisors on Science and Technology. (2012). 'Realizing the full potential of government-held spectrum to spur economic growth.' Available at: http://www.whitehouse.gov/sites/default/files/microsites/ostp/pcast_spectrum_report_final_july_20_2012.pdf. Accessed on 7 February 2013.

3

Conceptual Issues in the Allocation and Assignment of Spectrum

The ideas of economists and political philosophers, both when they are right and when they are wrong, are more powerful than is commonly understood. Indeed, the world is ruled by little else. Practical men, who believe themselves to be quite exempt from any intellectual influences, are usually slaves of some defunct economist.
—John Maynard Keynes in *The General Theory of Employment, Interest and Money* (1936), Chapter 24 'Concluding Notes'.

Introduction

The allocation and assignment of spectrum falls under three broad frameworks. In *command and control* many of the following decisions are administratively undertaken: spectrum use, choice of technology, number of blocks, identifying individual users, and the price paid; in *flexible use*, the job of spectrum managers ends at carving out blocks, and then auctioning them with rules to prevent interference; in *spectrum commons*, spectrum managers provide unrestricted, free access to users for specific blocks of spectrum with certain rules on the types of uses in place.

The spectrum-allocation process is a multi-level international process headed by the International Telecommunication Union (ITU).

The process involves many considerations other than the purely economic constraints of band availability, the pulls and pressures of standards bodies and member countries, and the need to facilitate progress to higher generation technologies. Nevertheless, it is important to understand some of the economic principles which the process of allocation must take into account. The phenomenon of adjacent spectrum bands like the 3 GHz and the 3.4 GHz commanding very different values in the marketplace indicates that the process of allocation may need to pay further attention to such considerations.[1]

Below the ITU are regional bodies and national spectrum managers who consider ITU's recommendations on spectrum use in the light of their priorities, finalize their frequency use plans, carve out spectrum blocks for individual users, decide on a method of selecting users, and then make further decisions on technologies to be adopted, the identification of users, and prices paid (if the auction mechanism is not being used).

In the context of varying degrees of command and control being the most widely used framework for spectrum management in different parts of the world, this chapter highlights some economic principles underlying decisions related to:

* Use to which spectrum is to be put
* Technology standards to be used
* The number of licensees[2]
* Administrative price

The choice of the mechanism to select users, in particular the design of the auction for spectrum and the most widely accepted method of spectrum assignment, is discussed in the next chapter. Issues related to the spectrum commons are also reserved for a separate discussion.

[1] Some believe that such phenomena undercut the very premise that the allocation of spectrum to different uses needs to be an administratively managed process (Cave et al. 2007).
[2] In some auction formats the number of licensees can emerge endogenously from the auction.

Determining Spectrum Use

The economic principle underlying the assignment of a spectrum band to different uses is the equalization of value in each use. The value of spectrum in a particular use is the number of units of a complementary input, for instance, labour or base transceiver stations, saved by an extra unit of spectrum at the margin multiplied by the price of the complementary input.

Table 3.1 provides a simple numerical example. For ease of understanding, one can think of spectrum 1 as a 700 MHz spectrum (traditionally used for broadcasting), and spectrum 2 as an 800 MHz spectrum (traditionally used for mobile communication); use 1 as broadcasting and use 2 as mobile communications.

Spectrum 1 has more value in use 2 and spectrum 2 in use 1. So more of spectrum 1 should be moved to use 2 and spectrum 2 to use 1. As more spectrum is allocated to a particular use its value in that use reduces due to the phenomenon of diminishing returns. A reduction in the value of one type of spectrum in a particular use is also caused by allocating more units of the other type of spectrum to that use. The converse is also true. Consequently, as we move spectrum 1 from use 2 to use 1 and spectrum 2 from use 1 to use 2, the value of a particular spectrum band in different uses will tend to equalize. Efficiency is achieved when the values are exactly equal as shown in Table 3.2. The price per unit of different spectrums should be set at

Table 3.1 Unequal values in different uses

	Spectrum 1	Spectrum 2
Use 1	100	150
Use 2	200	125

Table 3.2 Equal values in different uses after re-allocation

	Spectrum 1	Spectrum 2
Use 1	150	130
Use 2	150	130

the respective value accruing from the spectrums. This is referred to as the Smith-NERA method of administered pricing (Doyle 2004).

Note that different spectrums may not have the same value because they have different propagation characteristics or their associated ecosystems may be differently developed.

Sometimes when one spectrum is much better than the other in a particular use, then a boundary solution may emerge. In Table 3.3, even when the whole of spectrum 1 available is allocated to use 2, and the whole of spectrum 2 to use 1, the value in use 2 exceeds the value in use 1. Then efficiency considerations merit that spectrum 1 should be exclusively used in use 2. Similarly a boundary condition may emerge for spectrum 2 as well. The prices of the two spectrums should be equal to their value in the use to which they are allocated.

However, it must be noted that the economic principles constitute only one set of considerations for determining spectrum use. In Chapter 1 the multi-dimensional pressures brought to bear at a multilateral, regional, and national level on the allocation process have been highlighted. The claims of defence and the government always seem to carry a weight over and above commercial considerations. The freeing up of bands for unlicenced use triggers another set of stakeholder reactions. The advocates of flexible use argue that moving from a command and control paradigm to flexible use will eliminate a lot of the inefficiency and waste from the allocation process.

Choosing Standards

The phenomenon of network externalities relates to the increasing benefits to each member of a network from an increase in the number of members of the network. There are network externalities associated with telecommunications standards. These relate to the inter-operability of handsets between systems using different

Table 3.3 Exclusive use of spectrum band for single use

	Spectrum 1	Spectrum 2
Use 1	100	**150**
Use 2	**200**	125

standards, and the choices available to a subscriber using a specific standard. The presence of a compatible network increases the ease of roaming as subscribers do not have to switch between handsets. Further having one standard for the whole market allows a greater number of operators and thus increases consumer choice.

Shapiro and Varian (1999) argue that network externalities in the cellular mobile industry are 'strong, but not overwhelming'. For example, even if consumers are locked in into one system, they can switch to other systems at a discount in exchange for signing service contracts. They conclude that the market is not especially prone to tipping. And indeed, in none of the cases where competition between systems was allowed was there a system that eventually cornered the market fully and became the de facto standard (for example, the US digital cellular market still supports three systems).

In markets without network effects, allowing multiple competing systems seems to be unambiguously desirable. In contrast, in markets with network externalities there are both advantages and disadvantages of having multiple systems rather than a single standard. Supporters of a single standard argue that network externalities are realized faster and technological uncertainty among consumers is reduced. Advocates of competition between standards point out that a decentralized approach is the best guarantee for promoting technological progress for developing even better technological systems; it also reduces the risk of being locked in into an inferior technology promoted by the government. A counter-argument is that free markets may also lead to locking in into inferior outcomes due to path dependence (Liebowitz and Margolis 1995), thereby necessitating government intervention to cope with this negative externality.

During the analogue period, the European countries followed an uncoordinated approach, which resulted in a proliferation of systems. Some of these systems did not even find an application outside the home country. As a result, pan-European roaming possibilities were mostly limited. Notable exceptions were the Scandinavian countries whose coordinating efforts led to the NMT system with the possibility of roaming across Scandinavia.

In contrast to Europe, the US introduced a single standard, the AMPS system, with nationwide roaming possibilities (Gandal et al. 2003). This single standard promoted the consolidation of the industry where most of the cellular operators not related to the Baby Bells were acquired or merged into nationwide operators. The AMPS system also became the most widely used system in Central and Southern America.

When the digital technology emerged, the demand for a single standard in Europe was strong. The launch of digital cellular coincided with the effort to complete the European Union (EU) single market. Ensuring a level playing field for all competitors, as entailed by a common standard, was perceived to contribute to this policy goal. The European Commission spent considerable efforts in following a coordinated approach this time in collaboration with state-owned telecommunications operators and the equipment industry. The coordinating efforts resulted in the introduction of the GSM system as the single European standard.

In the US the desire for a single digital standard was not so strong. AMPS had already set a standard for the analogue technology and the main consideration was backward compatibility so that existing analogue users were not stranded. The result was three different competing systems. The first system, DAMPS, offered backward compatibility with AMPS. The second system was the GSM system, which had the advantage of compatibility with the European standard, and had become the most popular system worldwide. The third competing system was based on the new CDMA technology.

Of the 118 countries that adopted an analogue cellular system, 105 opted for a single standard and 13 for competing standards. A similar picture obtains for the countries that adopted a digital system. Of the 87 countries, 79 opted for a single standard and 8 for competing standards. Thus, there is a fairly constant fraction of countries (about 10 per cent) that adopts multiple systems (Gruber and Verboven 2001). For both analogue and digital technologies there is a single standard during the first years. While multiple systems of analogue technology start to appear only after 8 years, this already happens after two years with digital technology. Using a panel data set of annual data from 140 countries from 1981 to 1997 on mobile diffusion with single or competing standards, covering both the analog and digital phases, Gruber and Verboven (2001) show that competition between analogue technological systems significantly slowed down the growth in mobile diffusion, by a factor of about 5–15 per cent. This finding is consistent with the hypothesis that setting a standard makes consumers more eager to adopt a new technology in the presence of network effects. In contrast, the presence of competition between digital technological systems did not significantly affect the growth in mobile diffusion. The latter finding contrasts with a common view that installing the single digital GSM standard in Europe was responsible for the high mobile penetration levels.

The empirical results then indicate that the disadvantages of allowing competing systems were dominant during the analogue area, and were balanced by the advantages during the digital area. This confirms the analysis on digital systems competition advanced by Shapiro and Varian (1999), who argue that the decentralized approach followed in the US gave the innovative but controversial CDMA technology (in terms of capacity advantages) a chance to develop.

In 3G systems the deciding factor in the competition between the family of UMTS systems (that includes WCDMA) and CDMA 2000 was the compatibility of UMTS with the installed GSM systems and the incorporation in WCDMA of some of the fundamental technological advances of CDMA. Though the CDMA family (EVDO Rev A and Rev B) and WiMax provides better spectral efficiency than WCDMA, over 90 per cent of the 3G networks are WCDMA/HSPA.

In 4G, the GSM Association consisting of all GSM and UMTS equipment vendors and operators advocated and promoted FD-LTE as the natural progression path, providing adequate compatibility to existing 2G and 3G network systems. The TD-LTE promoted by China Mobile and Qualcomm however has fewer takers. The number of user devices for TD-LTE is just a fraction (about 20 per cent) of the FD-LTE devices available (GSA 2013).

Finally, today the possibility of 'flexible use' is under discussion where the government does not specify any technology and leaves it to the choice of the licensee with technology dependant property rights specified in advance to check interference. From the point of view of the regulator, the advantage of specifying some technology for a band of spectrum over 'flexible use' is that it allows the government to precisely model the interactions between neighbouring bands and specifying property rights. However, property rights' contingent on the combination of technologies used by neighbouring bands can also be specified.

Number of Blocks

A commonly held view is that competition is the most effective market structure for ensuring low prices and high quality. However, in industries such as telecom services and electricity distribution, economies of scale and scope are large enough to warrant low levels of competition, even monopolies, to minimize unit costs. Telecommunication carriers face huge initial costs, including, for example laying down copper lines from the central office to each subscriber location in case of basic fixed line services, constructing cell sites and base transceiver stations (BTSs) in case of mobile services, and laying optic fibre cables to inter-connect their access networks to backbone networks.

In contrast, the marginal cost of providing services to each additional customer, once the network is operational, is often negligible in comparison. Given the enormous fixed costs and negligible marginal costs, a carrier's long-run average costs within the defined geographical area may well decline with every increase in the size of the network. In other words, it is often cheaper for an operator to provide services to the one-millionth customer than to the one-thousandth customer (for details on economies of scale and scope in telecommunications, see Nuechterlein and Weiser 2005; Prasad and Sridhar 2008, 2009).

In the case of mobile services, economies of scale are driven by 'trunking efficiency' and wideband modulation schemes feasible on contiguous bands of spectrum. As the amount of spectrum deployed increases, the capacity of a network to carry traffic increases in a greater proportion than the proportion of increase in spectrum. In addition there are operational efficiencies related to general, administrative, and marketing costs.

Further, one must consider that there is a minimum feasible block size necessary for rolling out wireless services. This minimum size depends on the service in question and the technology deployed. For instance, in CDMA the minimum block size is 1.25 MHz. Given the total spectrum available, this translates into an upper cap on the total number of operators.

In the early days of mobile communications there was usually only enough spectrum for one or two operators. The capacities of the networks were also quite limited. This is reflected in the competitive choices made by national regulators. Of the 118 countries that adopted an analogue cellular system, 88 countries chose a monopoly (of which 83 had a single standard and 5 had multiple systems) and 30 countries chose an oligopoly. This relationship was reversed for the digital technology which exhibited far higher capacity than the analog technology. Of the 87 countries, only 39 had a monopoly (Gruber and Verboven 2001).Today, however, the upper cap in certain bands is much larger than the number of operators that would minimize unit costs in the presence of economies of scale.

Allowing a few firms to dominate a market would lead to greater efficiency in production, but at the risk of increased mark-ups over marginal cost on account of market power (Sridhar and Prasad 2011). On the other hand letting in the maximum feasible number of operators and allowing the forces of competition to decide how many survive (subject to a floor on the number of operators) may not be advisable given the large upper bound, the relatively low number of operators observed in the steady state in most countries, and the time taken for mergers and acquisitions (M&A) activity.

In the telecom industry there is a special reason why M&A is slow. To eliminate non-serious operators and for preventing speculative activity, the licence imposes roll-out obligations without the possibility of early exit. Therefore, mergers can only take place after multiple roll-outs have already been affected. This leads to a tremendous wastage of resources. Thus the licensor has to control the number of companies and cannot solely rely on the M&A environment to do so.

Baumol (1982) introduced the notion of contestability as a more fundamental conception of the competitiveness of an industry compared to conceptions based on market share. An industry is said to be perfectly contestable if three conditions are satisfied: firstly, new firms face no disadvantage with respect to existing firms; secondly, there are zero sunk costs; and thirdly, the entry lag is less than the price adjustment lag. The mobile industry is not perfectly contestable but in most contexts, there exist several avenues for increasing contestability. These include adopting mobile number portability, the licensing of mobile virtual network operators, the facilitation of spectrum and tower sharing, and allowing the entry of new operators through the release of new blocks of spectrum aligned with advanced technologies.

An attempt should be made to increase competitiveness without overly fragmenting the market. Although competition can unleash innovation, beyond a certain level it may degenerate into a price war and kill innovation in business models, technology, and operations.

Finally, there must be a facilitative M&A environment that allows market activity to determine the optimal number of operators. In mobile telecommunications the rules for M&A must specify a cap on spectrum as well as market share since an entity with an unacceptably large share of spectrum could gain an unacceptably large market share in the future. However, the caps on market share and spectrum should recognize the importance of allowing economies of scale to be reaped and of using all avenues for increasing contestability to decrease market power. Important issues also relate to whether the spectrum cap should be applied as a fraction of the total aggregated spectrum (allocated to different generations of technologies), or for kind of spectrum separately. In a liberalized regime where technology is not regulated by the government, the former may be appropriate. Similarly, the determination of the market share could use revenues in each service—voice, data, video, gaming, separately or as a whole. The answer to this question depends on the prevailing mode of delivering services to a customer. In a situation where most customers consume both data and voice, separating the two revenue streams may be inappropriate and even infeasible. The features of such M&A environments are discussed later in the context of the Indian telecom industry.

There has been some discussion on the timing of entry of new operators in the context of switchover costs in the mobile industry. Under simultaneous entry, firms maintain more or less symmetric market shares. This gives them incentives to compete more softly. In contrast, under sequential entry the second entrant starts from a low market share and needs to compete vigorously to obtain a part of the incumbent's installed base, which induces further aggressive responses by the incumbent. In the case of sequential entry, incumbents also have an incentive to pre-empt entry in the period prior to actual entry. This maybe done, for example, by charging low prices or following aggressive marketing campaigns. Gruber and Verboven (2001) show that the impact of competition on the diffusion of mobile telephony was significantly stronger when entry was introduced sequentially than when it was introduced simultaneously.

Setting the Spectrum Price

While auctions have become the preferred method of assigning spectrum in some countries, in many countries methods other than an auction continue to be used. Prices for spectrum are set based on its estimated market value and on the perceived impact of the price of spectrum on the price of telecommunications services and the speed of diffusion of telephony and the Internet. The relation between spectrum price and the price of service is a matter of some controversy.

One view is that spectrum is a sunk cost and hence does not affect market prices (Prasad and Sridhar 2011). Another view holds that spectrum prices do impact market prices and hence spectrum for basic services should be priced below value.

In a static economic model, the price of a good or a service depends only on market competition, and not on sunk costs like the cost of spectrum. However in the real world, sunk costs can impact market prices through three channels (Prasad 2011). First, they can impact the amount of R&D expenditure and thereby the innovation and competition in the industry.

To gain some perspective on this point, let us start with an episode from a different industry. There are many similarities between pharmaceuticals and mobile telecom services in that both industries involve high fixed costs and negligible marginal costs. Many years ago drug companies in the US protested against regulation that required drugs to be tested for their suitability for children with the argument that the regulation would increase pre-launch costs and thereby the price of drugs. An economist wrote an article in the *Wall Street Journal* pointing out the fallacy of their position: since R&D on already discovered drugs was a sunk cost, in other words a cost that could not be transferred, it would not affect the pricing decision.

The economist, while correct, did not tell the whole story. For while an increase in R&D costs already incurred was a sunk cost, future R&D was not. It was a cost whose level would be determined by the factory on the basis of its calculation of profit potential. An increase in the R&D cost structure of factories might inhibit future R&D leading to less competition in the market for already developed drugs and higher prices (or lower price reductions) than those that would have prevailed in the absence of the proposed regulation.

Further, high sunk costs can trigger exit of firms from the industry. The very real prospect of a serious player like Uninor leaving the Indian telecom industry on account of the hike in the spectrum fees is an instance of such a phenomenon. Exit reduces competition and can raise prices.

Third, if capital expenditures are not simultaneously but sequentially determined then high sunk costs can reduce subsequent capital outlays. For example, in telecommunications a high cost of spectrum can reduce investments in physical infrastructure leading to lower network quality and lower competition. Of course, decreases in sunk costs that merely add a sliver to profits are unlikely to have an impact on prices. But in a nascent market with liberal entry conditions, sunk costs significantly below value can lead to higher innovation, higher competition, and higher investment in complementary capital than substantially higher sunk costs, as firms compete aggressively to establish dominant positions.

Many countries like the UK, France, and the US have taken the view that spectrum for voice services, that is, 2G spectrum, needs to be subsidized while spectrum for higher value services, that is, 3G and 4G spectrum needs to be priced at value, with the value being determined in an auction.

The lower revenues from upfront charges for spectrum in some instances sought to be recovered through high spectrum usage charges, that is, a revenue share going to the government. Usage charges are not a sunk cost, but a variable cost like a sales tax.

The theory of optimal taxation (Mirlees 1986) advocates against introducing distortions in input markets through indirect taxes or subsidies. Instead it recommends that if indirect taxes or subsidies are to be used then these should be in the market for final goods and services. This approach limits the distortionary effect of government intervention and prevents the impact from cascading into the economy.

Like all such taxes, usage charges can have the effect of increasing prices. Further, sometimes spectrum usage charges increase at higher holdings of spectrum. This is the case in India. This makes them more distortionary than the ordinary sales tax, even one which increases with increasing revenue, because they penalize increases in revenue occasioned by an increase in the usage of only one of the inputs—spectrum. Hence, the production decision is biased in the direction of increased use of physical infrastructure.

Sometimes in the early stages of an industry, generating tax revenue through usage charges may be appropriate. If there is uncertainty regarding the future of the industry then it is better to have easy terms in the form of low upfront costs and higher revenue collection from those who do well in the market.

In some cases the government aims to maximize revenue through high upfront fees, but this may be a short-sighted way of looking at the problem. Future revenues through the diffusion of services also need to be taken into account. In general, governments aim to balance the goal of revenue maximization with rapid diffusion.

Conclusion

This chapter discussed some important issues that need to be addressed in creating exclusive licensing regimes for spectrum. These include the allocation of the use for different bands of spectrum, the choice of technology, the determination of the number of spectrum blocks, and the choice of the administered price (in case comparative hearings are used for assigning spectrum). As acceptance for auctions as a method of allocation becomes widespread one of the most important issues is the design of spectrum auctions and the design of secondary markets. These are taken up in a separate chapter.

References

Baumol, W. (1982). 'Contestable markets: An uprising in the theory of industry structure', *American Economic Review* 72(1): 1–15.

Cave, Martin, Chris Doyle, and William Webb. (2007). *Essentials of Modern Spectrum Management*. Cambridge, UK: Cambridge University Press.

Doyle, Chris. (2004). 'The economics of pricing radio spectrum', Department of Economics, University of Warwick 16. Available at: http://userpage. fuberlin.de/~jmueller/its/conf/berlin04/Papers/Doyle.pdf.

Gandal, Neil, David Salant, and Leonard Waverman. (2003). 'Standards in wireless telephone networks', *Telecommunications Policy* 27(5): 325–32.

Global mobile Suppliers Association (GSA). (2013). *LTE Ecosystem Wall Chart*. Available at: www.gsacom.com. Accessed on 10 April 2013.

Gruber, Harald, and Frank Verboven. (2001). 'The evolution of markets under entry and standards regulation—the case of global mobile tele-communications', *International Journal of Industrial Organization* 19(7): 1189–212.

Liebowitz, Stan J., and Stephen E. Margolis. (1995). 'Path dependence, lock-in, and history', *Journal of Law, Economics, & Organization* 11(1): 205–26.

Mirlees, James A. (1986). 'The theory of optimal taxation', *Handbook of Mathematical Economics* 3: 1197–249.

Prasad, R. (2011a). *Digital Crossroads*. Cambridge, MA: MIT Press.

———. (2011b). '2008 telecom licensing policy: Conceptual issues', *Economic and Political Weekly* XLVI (53).

———. (2011c). '2008 licensing policy: Conceptual issues', *Economic and Political Weekly* XLV (53).

Prasad, R., and V. Sridhar. (2008). 'Optimal number of mobile service providers in India: Trade off between efficiency and competition', *International Journal of Business Data Communications and Networking* 4(3): 69–81.

———. (2009). 'Allocative efficiency of the mobile industry in India and its implications for spectrum policy', *Telecommunications Policy* 33(9): 521–33.

Shapiro, C., and H.R. Varian. (1999). *Information Rules. A Strategic Guide to the Network Economy*. Boston: Harvard Business School Press.

Sridhar, Varadharajan, and Rohit Prasad. (2011). 'Towards a new policy framework for spectrum management in India', *Telecommunications Policy* 35(2): 172–84.

4

Auctions

In the movies, art auctions look like high-end carnival acts, with a sleazy barker/auctioneer keeping up a patter of 'I have $10 million, do I hear 20?' and hapless innocents accidentally holding up paddles and bidding on works they can't remotely afford. But at Sotheby's and Christie's, the only accurate part of that image is the paddles.

—Sean Rocha, writer and photographer[1]

Introduction

After allocating spectrum to different uses, deciding on the choice of standards, and the number of spectrum blocks, the method of assigning spectrum to different individual users must be chosen. Prior to assigning spectrum by auction, national spectrum managers, including the Federal Communications Commission (FCC), used administrative hearings (sometimes referred to as 'beauty contests') to choose individual licensees. These hearings evaluated the technological and financial fitness of applicants in a process involving lawyers, economists, and engineers and lasting several months. Lawmakers in the US worried that such a procedure would slow

[1] Available at: http://www.slate.com/articles/news_and_politics/explainer/2004/05/how_does_a_sothebys_auction_work.html. Accessed on: 20 December 2013.

down the introduction of new technologies and wanted a speedier and more transparent process. They therefore passed legislation that required that cellular licences be allocated by lottery. Specifically, all applicants who could provide documentation that they had the funding to build out a cellular business would be placed in a pool. The government would choose from among this applicant pool randomly. The large number of applicants for the lottery, many unqualified to build networks, and the long time required on secondary markets for spectrum to reach the hands of those willing and able to make use of it, resulted in lotteries also being eschewed. Auctions became the preferred method of spectrum assignment in the US and several other countries with users being given the right to exclusively transmit signals in a certain frequency range for specified services, in a given geographical area, for a stipulated period of time.

A great deal of attention has been paid to the appropriate design of auctions that will achieve the goals of spectrum policy. But much of the success of spectrum auctions depends on factors other than auction design, factors related to the overall paradigm of spectrum management—the release of adequate amounts of spectrum to avoid 'artificial scarcity', the manner in which property rights are defined—including rules on technology and service, the choice of spectrum blocks, the upper cap on spectrum holding, and the functioning of secondary markets.

In general, auctions are considered a preferred manner of allocating goods because they are believed to select a user who is best qualified to use the good, as indicated by the ability to bid the highest amount. Further, auctions allow the price of a good to be determined through a process that incorporates information that may be widely dispersed among different market players, instead of relying on an administrative process. The auction process is transparent, less subject to misuse of discretionary powers, and competitive. It can be designed to terminate fairly rapidly.

The objectives of an auction may be contravened by the same factors that cause failure in markets as a whole: the presence of externalities, market power and collusion on the buyer side, and asymmetric information between buyers and sellers. Another important challenge to efficiency stems from the uncertainty in nascent industries in the context of 'incomplete financial markets' that makes risk planning inefficient. This uncertainty may result in a winner's curse, a phenomenon where the winner of an auction pays more than the value of the object. Further, auctions are no better guarantees of addressing concerns of equity than any other market.

The auction of spectrum, a publicly owned and highly valuable resource, is a means of generating revenue for the government, and must ensure the avoidance of undue enrichment of private players at the expense of the exchequer. The process aims at an optimal level of competition between market players and the expeditious deployment of new technologies.

The design of spectrum auctions must take into account the specific characteristics of spectrum as an economic good. Spectrum blocks in a particular location are partly substitutable; blocks in different locations are partly complementary due to subscribers' need to roam, and partly substitutable due to the possibility of sub-national operators substituting similar regions in order to achieve their business goals; and spectrum blocks exhibit trunking efficiency, that is, a disproportionate increase in capacity for a given increase in the size of spectrum blocks.

Auction Formats

First- and second-price auctions: In a first price auction, the highest bidder gets the object at a price equal to his bid amount. In a second price auction, the highest bidder gets the object at a price equal to the bid of the second highest bidder.

Static and dynamic auctions: In static or sealed bid auctions, bidders make a single bid without observing others' bids. The physical metaphor is that of bidders placing their bids in a sealed envelope and handing them to the auctioneer. In a dynamic auction, bidders can observe others' bids, for instance, in an open outcry or ascending auction where bidding starts from a low amount and moves up. The descending auction where the auctioneer sets a very high price and continuously lowers it till a bidder calls out to buy is also said to be dynamic, since bidders can make strategies contingent on no other bidder stopping the auction till a certain point.

Combining the two aspects of auctions we arrive at four standard types of auctions that are used for the allocation of a single item:

- First-price sealed-bid auctions in which bidders place their bids in sealed envelopes and hand them to the auctioneer. The envelopes are opened and the individual with the highest bid wins, paying a price equal to the exact amount that he or she bid. This is also known as a simultaneous or static first-price auction.

- Second-price sealed-bid auctions (Vickrey auctions) in which bidders place their bids in sealed envelopes and hand them to the auctioneer. The envelopes are opened and the individual with the highest bid wins, paying a price equal to the exact amount of the second highest bid. This is also known as a simultaneous or static second-price auction.
- Open ascending-bid auctions in which the auction starts at a low price and moves up with bidders dropping out as the price becomes too high for them. The price may be raised either by 'open outcry' of the bidders (English auction), or by steady increases announced by the auctioneer (Japanese or 'clock' auction). The process continues till only one bidder remains who wins the auction at the current price. This is a dynamic second-price auction (since the final price is the price at which the second highest bidder in the fray drops out).
- Open descending-bid auctions (Dutch auctions) in which the auctioneer announces a very high price and progressively lowers it until a bidder indicates that she is prepared to buy. She is declared the winner and pays the bid price. This is a dynamic first-price auction (since the final price is the price the winning bidder is prepared to pay).

The four canonical auction formats are depicted in Table 4.1.

The first-price sealed-bid auction and descending auction are strategically equivalent. In a descending auction, the bidder has to choose a price to bid conditional on no one else having called out. Hence the auction, for all practical purposes, concludes without

Table 4.1   Auction formats

	Static	Dynamic
First-Price	First-price sealed-bid/static auction	Dutch Auction
Second-Price	Second-price sealed-bid/static auction/Vickrey Auction	English Auction/ Japanese Auction

observing the behaviour of others and is equivalent to the first-price sealed-bid auction.[2]

In multi-unit auctions, many units of a good, not necessarily identical, are put on the block. Each of the formats mentioned earlier may be used to auction them sequentially or simultaneously. It is also possible to run package auctions or combinatorial auctions where bidders bid for a combination of goods.

A Brief History of Spectrum Auctions

The first auction of radio spectrum took place in 1990 in New Zealand. Multiple television licences, each of 8 MHz were put on the block. A separate second-price sealed-bid auction was held for each licence with all bids for all licences being submitted at the same time. Given the complementarities of spectrum blocks in different regions, and the lack of clarity on the motivations of different operators, bidders had a difficult time deciding which regions to focus on. If they focused on regions that others also vied for they would end up paying more than they would if they chose less competitive regions. Each operator thus wanted to focus on blocks that the others did not focus on, except for national operators who bid everywhere. Given the sealed-bid auction format, there was no mechanism to resolve the coordination problem.

Some bidders bid for all regions but ended up winning only a few. Others bid for selected regions and ended up winning most of their targets. Given the nascent state of the mobile telecommunications market, the bids reflected an amalgamation of confusion, hope, and fear as much as they revealed as the operators' objective estimates of the value of spectrum. The licence prices ranged from NZ$400,000 to NZ$100,000. There was a public outcry when it was disclosed that a bidder was willing to pay NZ$2,371,000 for a licence but only paid

---

[2] The Revenue Equivalence Theorem states that under certain conditions all auction formats will yield the same revenue. This will appear to make an auction design (and this chapter) irrelevant. However, like many theorems in economics, the conclusion holds under assumptions that are not satisfied in the real world. The specific departure from the assumptions and the implications for auction design remain a fruitful area of research for game theorists.

NZ$401,000 because of the second-price design. There were two other New Zealand auctions where bidders paid far less than their bid price. In one auction, a bidder placed a bid for NZ$100,000 but paid only NZ$6 (the second-highest bid) and in another, a bidder placed a bid of NZ$7 million and paid only NZ$5,000. In 1991 and continuing through 1994, the New Zealand government reverted to first-price sealed-bid auctions.

The Swiss wireless-local-loop auction conducted in March 2000 illustrated the difficulties of a sequential sale. Three nationwide licences were sold in a sequence of ascending auctions. The first two licences were for a 28 MHz block; the third was twice as big (56 MHz). The first licence sold for 121 million francs, the second for 134 million francs, and the third (the large licence) sold for 55 million francs, only a fraction of the prices of the earlier licences.

In 1993, after suboptimal experiences with administrative hearings and lotteries, the US Congress authorized the licences of spectrum by auction. The FCC obtained advice from game theorists about the design of the auction format. They chose a novel auction design by choosing to auction all licences simultaneously in a dynamic auction process. This auction format, the simultaneous ascending auction, has become the workhorse of spectrum auctions the world over. As suggested by its name, it involves auctioning spectrum blocks across various regions simultaneously in a dynamic process involving bids being raised from a reserve price. Auctions using this format have been held across the world and yielded good outcomes: high growth of the industry and high revenues for the government.

In this format, many fine points relate to the manner in which the bids can be raised, the rules for stopping the auctions, and the required level of participation by the bidders through the entire process. These are elaborated upon after a discussion of certain conceptual aspects of auction design that will highlight the necessity for the finessing of auction rules.

Modelling Auctions

In game theory the canonical model of an auction is described by three parameters: (a) the set of bidders, (b) the set of bidding strategies available to each bidder, and (c) the payoffs corresponding to each combination of strategies (Klemperer 1999). A critical point of distinction between different auction models relates to their assumption about whether the object has 'private value', or 'common value'.

Private and common value auctions: In private value auctions: (a) each bidder has a personal value for the object which is independent of the value of other bidders, and (b) each bidder knows the value the object has for her. An auction of the private effects of a family member is often cited as an example of such an auction—one family member may have a high value for the objects of a favourite grandmother while another may have no such sentimental considerations.

In common value auctions all bidders have the same value for the object but different signals/information about it, and therefore different estimates. For example, in the auction of an oil lease different bidders have different geological information about the amount of oil contained. And yet, the value of the oil field will be the same for any winning bidder—it will be the actual amount of oil the field contains. In other words, the object has common value.

Most auctions are a hybrid of private and common value—bidders have their unique histories and strategies that predicate a private value for the object, but these private values are also tempered by technical and economic possibilities that are common to all.

Findings from Auction Theory

In an auction, every bidder wants to win the object at the minimum possible price that is no greater than the value that she ascribes to the object. In a first-price auction, the winner would rue the fact that she had to pay the value of her bid while she could have won even by bidding an amount fractionally higher than the second highest bidder. Unfortunately, bidders do not know that value at the time of making bids.

This feature of auctions constitutes one aspect of the two-fold nature of the winner's curse, on which there will be more discussion later. This leads bidders to try to 'shade' their bids, that is, bid slightly lower than their estimate of the value of the object. The phenomenon of bid shading was considered an issue by auction designers since one of the objectives of using the auction mechanism was to elicit the true valuation of the object.

Vickrey, a Professor at Columbia University, came up with a method guaranteed to make all bidders submit their true valuation in the context of private-value auctions (Vickrey 1961).

Vickrey's Truth Serum

Vickrey suggested that the winner of the auction should be the highest bidder, but the price paid should be the value bid by the second highest bidder. This is the second-price auction alluded to earlier. He argued that in *private-value second-price auctions*, no one will have any incentive to 'shade' (or inflate) their bids. If any bidder bids less than her value and wins, she will have to pay the second highest bid and receive the same payoff as in the case if she had bid her true value. However, by bidding less than her true value she runs the risk of losing an object which she would have liked to win in case the winning bid is less than her true value. Therefore, no bidder will bid less than the true value. In case the bidder bids more than her true value and wins, and if the second highest bidder is less than the first bidder's true value, the winner will be in the same position by bidding her true value. However, if the second highest bidder has bid higher than the winner's true value, then the winner will prefer to bid the true value and not incur the loss.

In sum, in a private-value second-price auction, bidding according to one's true value is the best course of action (a weakly dominant strategy in the parlance of game theory), and the bidder with the highest private value will win the object at a price equal to the second highest bid.

This is true irrespective of whether the second-price auction is a sealed bid or a dynamic/ascending auction. In a private-value ascending auction it is clearly a dominant strategy for a bidder to stay in the bidding till the price reaches her value. The person next to the last person standing will drop out when her value is reached and the person with the highest value will win at the price equal to the value of the second highest bidder.

Hence, an ascending auction is sometimes called an open second-price auction. However, exact equivalence only applies for a private values case. In the case of common values, bidders learn about the values of players when they quit the auctions and condition their behaviour on this information.

The Winner's Curse

The intuition of Vickrey's truth serum forms the basis of many auction designs using the second-price rule, like the spectrum auctions in New Zealand. However the truth-telling property of private-value, second-price auctions does not hold for common value auctions, even those using the second-price rule, as no one has complete information about the object and the actual value is indicated not just by a bidder's own signal but by the signals received by everyone.

In this context the Vickrey arguments break down. One can no longer assert that bidding the value indicated by one's signal will be better than bidding lower. Getting an object at the bid placed by the bidder with the second highest signal may not yield a profit for the highest bidder as the true value of the object may tend towards the value as perceived by the 'average bidder', the one who has obtained the signal which is the average of the signals received by everyone. Thus, we can no longer argue that by underbidding a bidder risks losing an object she could have got for a profit.

A key feature of bidding in auctions with common values is the winner's curse. Every bidder must reckon with the fact that if she wins it is because everyone bid less than her. Therefore, she has received the most optimistic signal about the true value. But the true value is also a function of the more pessimistic signals received by others. Thus, the winner is likely to be paying more than the value of the object, even if she is required to pay only the value of the second highest bid. Notice that the greater the number of bidders, the more inexorable the curse, since the probability that the losers are correct on average increases with the size of the bidding population. Empirical evidence indicates that winners of oil and gas drilling leases take substantial losses on their leases.

Awareness about the winner's curse leads everyone to adjust their bids downwards. Thus, a bidder has a double incentive to 'shade' her bid, first on account of wanting to bid as low as possible in order to win the auction, and second on account of avoiding the risk of paying more than the true value of the object.

One may think that the winner's curse is a boon for the auctioneer since it leads to high revenues. However, in the long run, auctioneers will gain nothing by beggaring their buyers since it leads to bankruptcy, discontinuities in business, and dampened commercial prospects all around. Auctioneers attempt to prevent a winner's curse by sharing as much information about the object being auctioned, screening bidders to exclude inexperienced ones who may bid unreasonably high, and adopting a dynamic auction format where bids are raised in an orderly manner, thus enabling market information to emerge for the benefit of all bidders. The risks of holding an auction with uncertainty are also mitigated by the presence of secondary markets for the object in question.

Beyond a threshold level of uncertainty, the auction mechanism may even be eschewed as it is likely to lead to a winner's curse. In such cases, objects can either be allocated as part of pilot projects given to research institutions, or assigned at highly subsidized prices with a revenue share agreement. The recipients of the subsidy should be chosen through an administrative evaluation process along technical and commercial dimensions.

Collusion

Collusion refers to surreptitious agreements or signals between bidders that reduce the upward movement of bidding and depress prices below the market value. Conceptually speaking, all one can assert is that collusion becomes less likely with a larger number of players. Eliminating collusion involves fine-tuning the details of the auction process. The risk of collusion is greater in a dynamic auction where bidding can take place over many days, even months, affording a window of opportunity for collusion to take place.

In a multi-unit auction, bidders have been known to ingeniously signal their intentions to each other using their bid amounts (Klemperer 2002). For instance, in the UK spectrum auctions held in 2000, bidders indicated their interests in specific regions by affixing the last two letters of the respective pin codes to their bids. Such collusion is prevented by mandating that bidders have to select from a drop-down menu of bid increments when deciding how much to bid in a subsequent round. An even more stringent requirement is that the bid increment is a fixed percentage increase over the previous round. One of the important instruments of guarding against collusion is setting a reserve price which represents a floor price for the object.

Reserve Price

A reserve price is the price at which bidding begins. If there is a transparent auction with enough competition among bidders, the final price will be largely independent of the reserve price (unless the reserve price is set too high in which case the auction may not result in a transaction). On the other hand, if there is only one bidder, or if there is collusion, then the final price will be close to the reserve price. In a collusive or insufficiently competitive auction, the reserve price protects the interests of the seller and ensures that the object is sold at a price which reflects its value. Auction theory prescribes setting a reserve price equal to the value that the seller could gain by retaining the object.

Since collusion can never be ruled out, the reserve price can never become entirely insignificant. However, in general, setting a reserve price equal to the administratively derived value of the object (in the hands of the buyer) and then inviting bids militates against the foundational premise that leads to the choice of the auction mechanism in preference to the administered price regime—the superior market information residing among market participants as compared to the auctioneer. It implies that administrative ignorance is unidirectional—only tending to underestimate the value of the object. The most important determinant of the reserve price is the level of competition in the auction. In spectrum auctions held since 2006, the reserve price was about 50 per cent of the final price.

Multi-unit Auctions

In multi-unit auctions, bidders are eligible to receive multiple units of an object, for example, sub-tracts of a large tract of land. These auctions are distinct from standard single-object auctions because of the linkages—complementarities and substitutability, between the objects.[3] For instance, the value of obtaining the entire tract of land may be more than the value of pieces of land acquired in isolation.

In such auctions, a sequential sale limits the information available to bidders and the manner in which they can respond to information. Bidders must guess what the prices will be in future auctions when determining bids in the current auction. Incorrect guesses may result in an inefficient assignment. A sequential auction also eliminates many strategies. A bidder cannot switch back to an earlier item if prices go too high in a later auction. Bidders may regret having purchased early at high prices, or not having purchased early at low prices. Guesswork about future auction outcomes makes strategies in sequential auctions complex and the outcomes less efficient. A sequential sale also exposes bidders to the *holdout problem*, the phenomenon of sellers gaining enormous bargaining power over buyers later in the process who have bid in earlier auctions in the hope of winning in later auctions and gaining synergies. For these reasons, complementary goods should be auctioned simultaneously rather than sequentially.

Simultaneous Ascending Auctions

The specific rules adopted in the simultaneous ascending auction, the most widely accepted method used for auctioning spectrum blocks across different regions, are now presented (Hoffman 2011). Drawing upon the findings of auction theory discussed earlier, the rationale for different rules is also discussed.

(1) Bidders simultaneously submit bids in each auction, round by round, until a round is reached in which no new bids are received in any auction. This gives bidders flexibility to target each region based on the progress of bidding in all the regions. For example, if bidding is in progress in only one remaining region with all the other regions closed, and if it goes very high, then some bidders

---

[3] Note that an auction in which many units of an object are auctioned, but each bidder is eligible to receive only one object is identical to the single-object auctions discussed earlier.

may want to reopen bidding in other regions which are substitutes for the last remaining region.

(2) After each round, the auctioneer announces a 'provisionally winning bidder' if that bidder had the highest bid in that round. A 'provisionally winning bidder' is responsible for paying his bid price unless he is outbid by another bidder in a later bidding round. In earlier versions of the simultaneous ascending auction, bidders could choose their bid increments. But this freedom became a signalling device in the hands of the bidders. Hence, they were restricted to selecting from a drop-down menu of possible increments, and more recently, to an automatic 5 or 10 per cent increment every round.

(3) Bid withdrawals were considered a useful tool to give flexibility to bidders to build a desired package of items. However, it became another tool for signalling as operators made and then withdrew bids in certain regions in order to signal their interests in other regions and induce like behaviour in their competitors. Currently, the possibilities for withdrawals are present but limited.

(4) In early auctions, bidders' identities and their respective bid amounts were public information. Given the need for operators to partner in order to provide national coverage, and for technologies used to be compatible, it was felt that revealing bidders' identities would enable more informed bidding. However, this transparency became a conduit for collusion. As a result, now bidders' identities are not revealed till the end of the auction, and only the provisionally winning bid amount is declared.

(5) In order to ensure that bidding in the rounds proceeds in an orderly manner and results in the revelation of market information for the benefit of all concerned, it is important to ensure that bidders act in a consistent manner through the auction. Inconsistency may be occasioned by not having thought through one's strategy or from a '*snake in the grass*' strategy whereby bidders lie quiet in early rounds and then surprise others by entering and bidding aggressively in later rounds. In order to achieve consistency of bidder behaviour, different regions are assigned 'bidding units' based on their business potential, and bidders are required to indicate the total number of bidding units they are interested in through an upfront payment. This determines their

'eligibility', that is, the upper cap on the number of bidding units that they can bid for in each round. In order to ensure regular participation in line with eligibility, there are 'activity rules' that stipulate that a bidder must score a minimum percentage of her eligibility in each round, that is, participate in a minimum number of auctions. This percentage increases as the rounds proceed. As the auction progresses and prices rise, the activity rule forces bidders to place binding pledges at the current prices or leave the auction.

(6) The winners in a simultaneous ascending auction, the only ones willing to pay the provisionally winning bid of the last round plus the increment stipulated by the auction designer, pay the bid of the second highest bidder plus the bid increment. The auction is therefore a second-price auction, with an additional 5 or 10 per cent added on, in line with the bid increment.

Simultaneous ascending auctions have led to orderly assignment of spectrum, high rates of growth of services, and many billions of dollars in revenue the world over. Some auctions like the auction for 3G licences in the UK in 2000 and the Indian 3G auction in 2010 are believed to have resulted in winning bids that prevented businesses from being viable. However, in general, the auctions are believed to have performed well. Although this auction format does not allow bidders to package licences into 'all or nothing' bids, bidders have the flexibility of moving among different band plans and collect a sufficient group of licences to develop usable business plans.

Package Auctions

In simultaneous ascending auctions, bidders bid for each individual item separately. The main advantage of this approach is simplicity (Cramton 2006)—the auction is easily implemented and understood. The disadvantage is the *exposure problem*. Although a simultaneous ascending auction allows bidders the flexibility to build desired aggregations of units, with individual bids, bidding for a synergistic combination can be risky. The bidder may fail to acquire key pieces of the desired combination, but pay prices expecting the synergistic gain. Alternatively, the bidder may be forced to bid beyond her valuation in order to secure the synergies and reduce its loss from being stuck with the individual pieces of the desired configuration of units. When the synergies between the various units are high, it may become necessary to use package or combinatorial auctions.

In package auctions, bidders are allowed to bid for combinations of objects or, in the case of spectrum, licences. While the total number of packages that a bidder can choose is enormous, the auction designers restrict[4] bidding to certain predetermined packages. For example, in the 700 MHz package auction held in the US in 2000, the bidders could choose one of 12 packages (FCC 2000). These packages need not be mutually exclusive. For example, a bidder could place a package bid for licences A and B, and simultaneously another package bid for licences B and C.

Allowing package bidding introduces a *threshold problem*—the difficulty that bidders for single licences (or smaller packages) that constitute a larger package may have in outbidding a single bidder on the larger package, even though the multiple bidders may value the sum of the parts more than the single bidder values the whole. This may occur because bidders for parts of a larger package each have an incentive to hold back in the hope that a bidder for another part will increase her bid sufficiently for the bids on the pieces collectively to beat the bid on the larger package.

The reserve price for any package is equal to the sum of the reserve prices of the individual licences. The provisional winning bid in any round is determined by computing the set of 'consistent bids', which will maximize the overall proceeds of the auction. Consistent bids are those bids which are mutually exclusive and pan the entire set of licences on the block.

As in the case of the SAA, activity rules are necessary to ensure that the bidders act in a consistent manner. The maximum eligibility of a bidder is determined in the same way as in the SAA—regions are given points based on revenue potential, a per point value is determined, and an upfront payment determines the maximum amount of points for which bidders are eligible to bid. A bidder has to bid on at least 50 per cent of their maximum eligibility in each round.

---

[4] With 'n' objects, the total number of packages is equal to $(2^n - 1)$.

A bidder who fails to meet the activity requirement in a given round will have her eligibility reduced for the next round to two times her activity in the current round.

A bid is considered 'active' if it is either a 'retained' bid[5] from the previous round or an accepted bid in the current round. The bidding units associated with licences on which a bidder was active, including retained bids, will count towards the bidder's activity. To account for the possibility of overlapping bids, which by definition cannot simultaneously be part of the winning set, a bidder's activity level in a round is the maximum number of bidding units that the bidder can win considering only the licences and packages on which the bidder is active, that is, counting the set of bids with the most bidding units in the case of mutually exclusive bids. For example, licence A has 10 bidding units associated with it; licence B, 20; and licence C, 20. If the only bids made by a bidder were on packages AB and BC, her activity would be 40 since AB and BC are mutually exclusive (that is, licence B is included in both packages, but can only be awarded as part of one package) and the package BC has more bidding units. Each bidder in the auction is provided a few activity rule waivers that may be used in any round during the course of the auction. An activity rule waiver applies to an entire round of bidding and not to a particular licence.

One effect of allowing mutually exclusive bids is that a bid does not necessarily have to be the highest bid on a particular package or licence in order for it to be a provisional winner. An example will illustrate this point: Bidder 1 places a bid of 50 on package A, and bidder 2 places a bid of 50 on package B. In the next round, bidder 1 places a bid of 100 on package B, which is mutually exclusive with her bid of 50 on package A from the previous round. If bidder 3 is allowed to bid 40 on package A, even though it is not higher than bidder 1's bid of 50, bidder 3 will become a provisional winner (assuming that these are the only bids). Bidder 3's bid of 40 on package A plus bidder 1's bid of 100 on package B totals 140, and this total is higher than bidder 1's bid of 50 on package A plus

---

[5] A 'retained' bid is defined as a provisionally winning bid or a bid that has the potential to become a provisionally winning bid because of changes in other bids in subsequent rounds.

bidder 2's bid of 50 on package B which totals only 100. We wish to encourage such bids. Moreover, bidder 3 may not have bid if she was required to beat bidder 1's bid of 50 on package A, which is not an efficient outcome.

Thus, unlike in the case of the SAA, the provisionally winning bid on a package could fall in the course of the auction. While this has the advantage of efficiency cited earlier, the risk of bids falling too low must be addressed by a suitably designed rule on bid increments. In the US auction for 700 MHz, the minimum accepted bid for any licence or package was the greater of: (a) the minimum opening bid; (b) the bidder's own previous high bid on that package plus x per cent, where the auctioneer specified the value of x in each round; and (c) the number of bidding units for the licence or package multiplied by the lowest $/bidding unit on any provisionally winning package in the last five rounds.

Under part (c) of the formula, the least expensive provisionally winning 'unit price' (the provisionally winning bid for a licence or package divided by the number of bidding units associated with the bid) for the five prior rounds is calculated. To perform this calculation, all of the provisionally winning bids for the five prior rounds are examined. Each of those provisionally winning bids is divided by the number of bidding units associated with it, to yield a 'unit price' for each provisionally winning bid. Finally, the lowest unit price of all of the provisionally winning bids (in other words, the lowest unit price that any bidder has bid for any provisional licence or package in the prior five rounds) is determined. To apply part (c) of the formula to a new bid, that lowest unit price is multiplied by the bidding units associated with the licence or package for which the bidder is bidding. It is possible, and indeed likely, that the lowest unit rate will come from a different licence or package than the one on which the bidder is bidding.

Part (c) of the formula essentially requires that bids on any licence or package be not too far from the provisionally winning bids; unless such a provision is included, bids might not become competitive without many rounds of bidding. Part (c) thereby facilitates bids that will overcome the threshold problem. By using the least expensive provisionally winning rate for any licence or package over the previous five rounds, auction designers have attempted to ensure that minimum accepted bids will not be too high, albeit with the loss of some simplicity in auction rules.

All licences remain open until two consecutive rounds have occurred in which no new bids are accepted (retained bids do not count as 'new bids'). After the second consecutive such round, bidding closes simultaneously on all licences. Thus, unless circumstances dictate otherwise, bidding remains open on all licences until bidding stops on every licence.

Conclusion

Auctions have been a successful method, and the simultaneous ascending auction a proven format for assigning spectrum rights. In conclusion, it is necessary to caution the government not to focus on the revenue generation possibilities of spectrum alone. Maximization of revenue can be achieved by releasing limited amounts of spectrum and assigning monopoly rights to the winning bidder. However, this will be at the cost of efficiency and technological progress. To recall the advice of old Polonious in Shakespeare's *Hamlet*, governments should give 'no unproportioned thought his act' when it comes to auction design.

References

Cramton, Peter. (2006). 'Simultaneous ascending auctions', in Peter Cramton, Yoav Shoham, and Richard Steinberg (eds.), *Combinatorial Auctions*, Chapter 4, pp. 99–114. USA: MIT Press.

Federal Communications Commission (FCC). (2000). *Procedures Implementing Package Bidding for Auction No. 31*. Available at: www.spectrum-exchange.com/files/da001486.doc. Accessed on 25 February 2013.

Hoffman, Karla. (2011). 'Spectrum auctions', in Jeff Kennington, Eli Olinick, and Dinesh Rajan (eds.), *Wireless Network Design: Optimization Models and Solution Procedures*, chapter 7, pp. 147–76. NY: Springer New York.

Klemperer, Paul. (1999). 'Auction theory: A guide to the literature', *Journal of Economic Surveys* 13(3): 227–86.

———. (2002). 'What really matters in auction design', *Journal of Economic Perspectives* 16(1): 169–89.

Vickrey, William. (1961). 'Counterspeculation, auctions, and competitive sealed tenders', *Journal of Finance* 16(1): 8–37.

5

Economic Models for Valuing Spectrum

> Nothing is more useful than water: but it will purchase scarce anything.... A diamond, on the contrary, has scarce any use-value in use; but a very great quantity of other goods may frequently be had in exchange for it.
>
> —Adam Smith, *The Wealth of Nations* (1776)

Introduction

It often becomes necessary to create economic models to value spectrum. For instance, an auctioneer requires some idea of the market value to set the reserve price. Bidders also need to estimate the value of spectrum in order to bid. One of the most common applications of spectrum valuation is in assigning spectrum to captive users—those who utilize spectrum for their internal operations and not commercial telecommunications services.

In many countries the predominant share of spectrum is allocated for such captive use. Further, governments have at various points in time chosen not to auction commercial spectrum but assigned it through an administrative process with administratively determined prices. In India, the 2G licence bundled with spectrum was assigned based on auctions in 1995, 2001, and 2012–14. Spectrum in 3G and BWA bands was auctioned in 2010. In 1999, the 2G licensees were migrated to a revenue share, and from 2001 to 2008, licences and spectrum were assigned on the basis of an administratively determined price. In all these cases, being able to value spectrum appropriately becomes critical. Indeed, the economic valuation of spectrum is important both in a market determined as well as an administrative process of assignment.

Some of the prevailing methods of spectrum valuation consist of benchmarking the current value to some proximate auction, or carrying out a calculation based on the capacity of a given spectrum block to carry signals. For instance, the Indian telecommunications regulator benchmarked the value of spectrum in the 1800 MHz band to the value of the spectrum in the 2100 MHz band discovered in the 2010 auction, after applying a correction based on the different propagation characteristics of the two bands (TRAI 2012). In the context of a captive spectrum the Indian regulator has been carrying out a calculation based on the quantity of bandwidth and the radius over which the spectrum rights are being assigned.

The benchmarking of values to proximate auctions is appropriate when the auctions in question have been held in the recent past and the spectrum rights being assigned are similar. When distant auctions are used for benchmarking, one must factor the change in revenue per megahertz of spectrum and the percentage increase in complementary inputs like Base Transceiver Stations (BTSs). In any case, the benchmarking exercise in such cases is not very satisfactory.

The methods used for valuing captive spectrum often focus on spectrum capabilities to the exclusion of the demand for spectrum. Adam Smith, in his seminal treatise *The Wealth of Nations* (1776) remarked on the difference between a product's value in use and its value in exchange. Water is extremely usefule but not valuable. The reverse is true of diamonds. While the growing scarcity of water threatens to make the comment out of date, the point about incorporating demand into economic valuation remains irrevocably pertinent. Spectrum has higher value in more dense, richer regions and this fact needs to be taken into account by administrative methods.

This chapter presents two different methodologies of estimating spectrum value. The 'bottom-up approach' determines the Net Present Value (NPV) of the cash flow that an operator will command by virtue of holding a block of spectrum over the licence period. The 'top-down approach', on the other hand, benchmarks the value of spectrum to the value of the physical infrastructure that the possession of 1 MHz of spectrum will substitute. Both methods are presented in the context of an application to the Indian case and draw on previous work (Prasad and Sridhar 2008, 2009; Prasad 2010; TRAI 2011).

As detailed elsewhere the Indian market is sub-divided into 22 licensed service areas or 'circles' categorized as metros, category A, category B, and category C based on revenue potential. The circles are often congruent with state boundaries. One such circle classified as category A is Tamil Nadu. Tamil Nadu is one of the advanced states in India with a per capita income of Rs. 72,993 (at 2013–14), a population of 7.21 crore, and an area of 1.3 lakh sq km. It comprises large rural areas that have 52 per cent of the state's population. The calculation is illustrated with an application to Tamil Nadu.

Estimating the Value of Spectrum Using the Cash Flow Method

The price per megahertz of a block of spectrum in Tamil Nadu is computed by determining the NPV over the licence period of 20 years of the cash flow that an operator in March 2010 will command by virtue of holding 6.2 MHz of spectrum. Note that while a new operator in India gets 4.4 MHz of spectrum, his operations are not stable. Hence, we use data corresponding to operators who have established themselves and graduated to holding 6.2 MHz of spectrum.

The cash flow accruing from the possession of a given block of spectrum is equal to the revenue earned from subscribers less the costs: the sum of the cost of the physical network,[1] spectrum charges, licence fees, and administrative, marketing and operating costs, that is:

$$Cash\, Flow = Revenue - (Network\, Cost + Spectrum\, Charges$$
$$+ Licence\, Fees + Administration, Marketing, \qquad (5.1)$$
$$\&\, Personnel\, Cost)$$

[1] At first glance it would seem natural to start directly with the average profits earned by firms holding 6.2 MHz or even 4.4 MHz. However, these profits are not available at a circle level. We therefore have to calculate these profits using data available, or through estimation.

The details of the method including the choice of the sample set of operators, the revenues, the cost of the physical network, spectrum charges and licence fees, and the administration, marketing, and operating costs are now presented.

Sample set of operators: Only data from operators who acquired spectrum in a circle a sufficient number of years in the past should be considered, as considering new entrants who have acquired spectrum but are yet to fully utilize it would artificially reduce the value of spectrum. In our computation we include operators who started operations on or before 2006, that is, four years in the past—Airtel, Vodafone, and Idea. We remove the public sector operators from the total spectrum held, as well as from the total subscribers serviced since their operations may not always reflect pure commercial considerations. In the exposition, all totals at a circle level should be taken to mean totals with respect to the sample set of operators unless otherwise mentioned.

Revenue: It is often not possible to use actual data to arrive at the revenue figure for a representative firm because the adjusted gross revenue (AGR) data at a circle level aggregates wireless and wireline access services. Hence, in our model, revenue is equal to the product of the number of wireless subscribers and the average revenue per user (ARPU) per annum.[2]

Assuming that operators with 6.2 MHz can command a subscriber base proportional to the amount of spectrum that they hold, the fair share of subscribers is equal to the proportion of spectrum held (6.2 MHz divided by the total spectrum assigned to the sample operators in that circle) multiplied by the total number of subscribers of the sample operators in that circle. We take the number of subscribers and the spectrum allocated in 2010 as the base for calculating the fair share.

In Tamil Nadu, the total subscriber base is 26.15 million, and the total spectrum held is 26.6 MHz. Thus, a fair share of subscribers of a representative operator with 6.2 MHz is 6.1 million. The ARPU is Rs. 174. Hence the annual revenue is Rs. 1272.66 crore.

Physical network: The cost of the physical network is equal to the cost of the BTSs and associated towers and the cost of the

---

[2] This only looks at mobile revenues.

core network, which includes transmission and switching. The cost of the BTSes is equal to the number of BTSes multiplied by the cost per BTS, including the rental and electricity costs associated with the physical infrastructure, while factoring the incidence of tower sharing observed in the market (Table 5.1).

Our aim is to estimate the average number of BTSs held by an operator with 6.2 MHz. If at least two operators in a circle with 6.2 MHz or below are not present, we take the BTS-spectrum ratio of the sample operators as a whole and fix the BTSes for our representative operator in a proportional fashion. If we do have two or more operators with 6.2 MHz or below we take their BTS-spectrum ratio alone in our calculations. From data at an all-India level there appears to be no correlation between the BTS-spectrum ratio and the quantity of spectrum held so a simple pro-ration, where necessary, appears to be justifiable. In Tamil Nadu the average number of BTSs obtained from such a calculation is 5,879.

The cost of a BTS as provided by TRAI is presented in Table 5.1.

The cost of the core network, as provided by industry sources, is around Rs. 500 per subscriber. Amortizing this over 20 years at 11 per cent, we get an annual cost of Rs. 53 per subscriber. Multiplying this by the number of subscribers we get the annual cost of the core network.

The total cost of the physical network in Tamil Nadu is thus Rs. 422.95 crore.

Spectrum charges, licence fees: We compute spectrum charges and licence fees according to the figures given in Tables 5.2 and 5.3 respectively.

Table 5.1   Cost of BTS

Amortization period in years	Rs. 20
Capital expense per BTS	Rs. 6,00,000.00
Rate of interest	Rs. 10%
Amortized capex	Rs. 64,068.89
Electricity and rental per year	Rs. 6,00,000.00
Total cost per year	Rs. 6,64,068.89

Source: TRAI (2011).

Table 5.2   Annual spectrum charges

Quantum of Spectrum Allotted	Annual 2G Spectrum Charges (as percentage of Adjusted Gross Revenue)	Annual 2G + 3G Spectrum Charges (as a percentage of Adjusted Gross Revenue)
For GSM licensees		
Up to 4.4 MHz	3	3
Up to 6.2 MHz	4	4
Up to 8.2 MHz	5	5
Up to 10.2 MHz	6	6
Up to 12.2 MHz	7	7
Up to 15.2 MHz	8	8
For CDMA licensees		
Up to 5 MHz	3	3
Up to 6.25 MHz	4	4
Up to 8.2 MHz	5	5
Up to 10.2 MHz	6	6
Up to 12.2 MHz	7	7
Up to 15.2 MHz	8	8

Source: Department of Telecommunications (2009).

Table 5.3   Licence fees for different categories of circles

Annual Licence Fees	2010–11 (Percentage)
Metro	10
Category A	9
Category B	7
Category C	6

Source: Department of Telecommunications (2009).

The spectrum charge is calculated from the spectrum charge percentage and the corresponding revenues; the licence fee is computed from the licence fee percentage and the corresponding revenues; and these are deducted from the total revenue. The total levy, that is, spectrum charges and licence fees in Tamil Nadu in Year 1 comes to 13 per cent, that is, 9 per cent for licence fees and 4 per cent for spectrum charges. Applied on a revenue of Rs. 1272.66 crore, the absolute levy equals Rs. 165.5 crore.

Administration, marketing, and operating costs: The total administration, marketing, and operating costs of operators as a percentage of revenue as presented in the accounting separation statements submitted by operators to TRAI as per statutory disclosure vary from 22 to 30 per cent. The percentage is lower for operators with a higher number of subscribers reflecting economies of scale. We take the percentage as 28 per cent for small operators, those with 6.2 MHz, and 22 per cent for larger operators. An operator holding 6.2 MHz, our representative firm in Tamil Nadu, incurs a cost of 28 per cent of AGR amounting to Rs. 356.3 crore. Deducting the cost of the network, the licence fees, spectrum charges, and general, marketing, and operating costs from the AGR gives us the cash flow accruing from holding a 6.2 MHz block of spectrum in Year 1. In Tamil Nadu this comes to Rs. 327.9 crore. The NPV over 20 years at 11 per cent, the weighted average cost of capital suggested by TRAI in its May 2010 recommendation (TRAI 2010) gives the value of 6.2 MHz. This is Rs. 2611.35 crore.

The price charged from the operator must allow a reasonable rate of return on investment, that is, on the price he pays for the spectrum. We fix the rate of return at 20 per cent. The value of spectrum less the NPV of the annual return, that is, 20 per cent of the price, over 20 years gives the price for 6.2 MHz for 20 years, that is:

$$Price = Value - NPV \ over \ 20 \ Years \ of \left(Price * 20\% \right) \quad (5.2)$$

Equation 5.3 allows us to compute the price of the spectrum. In Tamil Nadu this comes to Rs. 1322.9 crore for 6.2 MHz for 20 years. The price per MHz for 20 years is thus Rs. 213.4 crore. This price represents the weighted average of the price of 900 MHz and 1800 MHz spectrums where the weights are the proportions of 900 MHz and 1800 MHz spectrums being used in the circle in question, and where the price of 900 MHz spectrum is two times the price of 1800 MHz spectrum based on the propagation characteristics:

$$\begin{aligned}
&Percentage \ of \ 900 \ MHz * \left(Price \ of \ 1800 \ MHz * 2 \right) + \\
&\left(Percentage \ of \ 1800 \ MHz \right) * Price \ of \ 1800 \ MHz = \\
&Weighted \ Average \ of \ Price \ of \ 900 \ MHz \ and \ 1800 \ MHz
\end{aligned} \quad (5.3)$$

Table 5.4   Calculation of value of contracted spectrum in Tamil Nadu

No.	Field	Rs. crores
1	Revenue	1272.7
2	License fees	114.5
3	Spectrum charge	50.9
4	Core network cost	32.5
5	BTS cost	390.4
6	Operating cost	356.3
7	Annual cash flow (1–2–3–4–5–6)	327.9
8	NPV (cash flow) over 20 years (11% interest)	2611.4
9	Price of 6.2 MHz	1322.9
10	Price per MHz (blended 900 MHz and 1800 MHz)	213.4
11	Price per MHz 1800 MHz	136.3
12	Price for 4.4 MHz 1800 MHz	599.5

From Equation 5.3, knowing the percentage of 900 MHz and 1800 MHz, and the weighted average of the price of 900 MHz and 1800 MHz (derived earlier), we extract the price per megahertz of a 6.2 MHz block of spectrum in the 1800 MHz range. This comes to Rs. 136.26 crore (see Table 5.4 for the sample calculation for Tamil Nadu).

The values of 4.4 MHz of spectrum using the cash flow method in metros and category A circles is given in Table 5.5.

Table 5.5   Value of spectrum using the cash flow method

Circle	Estimated price of 4.4 MHz of 1800 MHz (Rs crores)
Delhi	483
Mumbai	329
Kolkata	164
Andhra Pradesh	485
Gujarat	473
Karnataka	432
Maharashtra	363
Tamil Nadu including Chennai	600

Operators with start-up spectrum face many challenges related to rolling out their networks, building a brand, and acquiring market presence. Operators with greater amount of spectrum and subscribers enjoy economies of scale in the form of lower administration, marketing, and operating costs. Hence, the value of the incremental spectrum is greater than that of start-up spectrum, and valuation of the incremental spectrum must be carried out separately using relevant parameters. The method of computation, that is, the cash flow model, however, remains the same.

Estimating the Value of Spectrum Using the Production Function Method

In the production function method, the value of spectrum is determined by the value of the physical infrastructure that 1 MHz of spectrum can substitute at the margin. This is akin to the Administered Incentive Pricing (AIP) schemes in the United Kingdom (Cave et al. 2007). In order to compute this value, we need to determine the relation between the quantity of spectrum and physical infrastructure used and the level of output of services, that is, the production function of mobile services.

The production function approach relies on specifying a functional form. The Cobb-Douglas function is widely used for estimating the statistical relationship between inputs and output. The production function is specified as:

$$X = Ay^{\beta} z^{\gamma} \tag{5.4}$$

where the dependent variable X refers to the mobile subscriber base, which is a proxy for minutes of use (MoU). The two factor inputs considered as the independent variables are: (a) allocated amount of spectrum that provides the required channel capacity for traffic (y), and (b) deployed infrastructure such as BTS (z) which provides connectivity to mobile handsets. The beta (β) and gamma (γ) values reflect the percentage change in subscriber base for a unit percentage increase in spectrum and BTS, respectively, and are the parameters to be estimated, besides A, which captures the magnitude of technical change. The major strengths of the Cobb-Douglas production function are its ease of use and its seemingly good empirical fit across many datasets (Miller 2008).

Our specification assumes that the two inputs, that is, spectrum and BTS can be substituted for each other over a certain range of output to service subscribers. The assumption of profit maximization implies that service providers will use an optimal mix of BTS and spectrum and that this optimal mix is determined by input prices. A higher charge for spectrum will induce service providers to substitute the less expensive BTS for spectrum to service the same subscriber base. The converse is also true.

A standard procedure for estimating the production function is to linearize it by taking logs on both sides. Thus, Equation 5.4 can be expressed as:

$$\ln X = \ln A + \beta \ln y + \gamma \ln z \qquad (5.5)$$

where β and γ measure the responsiveness of output to changes in levels of spectrum and BTSs, respectively, keeping the other input constant and are estimated using data for subscribers, BTSes, and amount of spectrum held by mature operators across the different categories of circles (the data that we use is described subsequently). For example, if $\beta = 1.15$, a 1 per cent increase in spectrum would lead to approximately a 1.15 per cent increase in the number of subscribers while maintaining the same number of BTSes. The estimated parameters of the production function are eventually used to derive the value of the 2G spectrum relying on the substitutability between BTSes and spectrum. If the service provider were to give up 1 unit of spectrum, he would need additional BTSes to be able to serve the same subscriber base. Since the price of BTSes is known, the value of the 2G spectrum can be derived as an opportunity cost, that is, savings in cost in terms of BTSes conserved by deploying an additional unit of spectrum.

The mathematics for the calculation makes use of the principle that at the optimum a service provider will allocate expenditure between the two inputs in such a manner that they yield the same marginal productivity per rupee spent.[3] The condition for optimality accordingly is given by:

$$\frac{MP_y}{P_Y} = \frac{MP_z}{P_Z} \qquad (5.6)$$

[3] This is, in fact, the solution to the firm's maximization problem in the face of a budget constraint. See, for example, Varian (1992).

where MP_y is the marginal productivity of spectrum and MP_z is the marginal productivity of BTSes. Deriving the marginal productivities from the functional form of the production function:

$$MP_y = \frac{\beta A y^B z^r}{y} \tag{5.7}$$

$$MP_z = \frac{\gamma A y^\beta z^r}{z} \tag{5.8}$$

the value of spectrum, denoted by P_y is derived as:

$$P_y = \frac{\beta z}{\gamma y} P_z \tag{5.9}$$

In Equation 5.9, P_z is the known price of a BTS, z is the number of BTSs deployed by the representative service provider holding 6.2 MHz and therefore known, y is the amount of spectrum held (6.2 MHz), and β and γ are estimated coefficients of the production function. The only unknown therefore is the price of spectrum which is calculated based on a combination of actual data and estimated coefficients of the production function.

A panel data set consisting of five category A circles for different GSM operators over the period 2007–10 has been used in the model. As in the case of the cash flow method, we do not include public sector operators in the sample because of their high spectrum allocation and unique operating constraints. We believe that this increases the reliability of the estimates in every circle. Finally, only mature operators, that is, those who have been in the market for at least 4–5 years are considered. Newer entrants are likely to focus on customer acquisition and network coverage making their BTS-spectrum trade-off very different from that of established operators.

Before carrying out the regression it is important to recall a technical attribute of spectrum which asserts that the load-carrying capacity of every megahertz beyond 3 MHz is about 3.75 times the load carrying capacity of each Megahertz up to 3 MHz. In order to homogenize the data on spectrum, we need to convert every block of spectrum into its corresponding quantity of effective spectrum, defined as spectrum beyond 3 MHz. If we fail to do this, given the variation in the holding of spectrum in different data points, we are likely to get a misleading estimate of the per megahertz value of the spectrum block of 4.4 MHz. For instance, the presence of a number of data points with holdings of 8 MHz of spectrum (each of which has 5 MHz of spectrum beyond 3 MHz) will give an overestimate of the per megahertz value of 4.4 MHz of spectrum.

We convert actual holdings of spectrum into units of effective spectrum as: a holding of x MHz consists of 3 MHz with capacity equal to $(1/3.75)$ of the capacity of effective spectrum, and $(x-3)$ MHz of effective spectrum. Hence, the total number of units of effective spectrum is given by:

$$\text{Number of units of effective spectrum} = (3/3.75 + x - 3)\,\text{MHz} = (x - 2.2)\,\text{MHz} \tag{5.10}$$

The regressions are carried out with these recalibrated spectrum holdings.

All coefficients are statistically significant (see Table 5.6) and the high R^2 inspires confidence in the overall significance of the model.[4]

It is important to note that the estimated coefficients are a product of both technology and market factors. For example, the high value of β could reflect the high density of subscribers as well as better spectrum management techniques.

The cost of BTSes for 2010 was estimated to be Rs. 6,64,069 per annum on the basis of data from TRAI.

The annual value for spectrum in Tamil Nadu, estimated on the basis of Equation 5.7, is Rs. 55.87 crore. Since spectrum is held by operators for 20 years, the annual value is thereafter converted to a 20

Table 5.6 Estimated parameters of the production function

Year		log spectrum	log BTS	R^2
	Coefficients	0.7	1.1	
	p-value	0.000	0.000	0.81

[4] The goodness of fit of estimation is given by the 'R^2' which is the variation in the subscribers that is explained by the variation in the two inputs. R^2 in our estimations is very good, above 80 per cent in each of the metros and category A (ideal fit being 100 per cent which is impossible).

year value by taking the NPV for 20 years at the prevailing SBI PLR for 2010. This comes to Rs. 444.93 crore.

As before, the weighted average of the price of 900 MHz and 1800 MHz spectrums is converted into the value of 1800 MHz spectrum as:

$$Percentage\ of\ 900\ MHz * \left(Price\ of\ 1800MHz * 2\right) +$$

$$\left(Percentage\ of\ 1800\ MHz\right) * Price\ of\ 1800\ MHz = \qquad (5.11)$$

$$Weighted\ Average\ of\ Price\ of\ 900\ MHz\ and\ 1800\ MHz$$

The percentage of 900 MHz spectrum in Tamil Nadu is 82.35. The value of a MHz of 1800 MHz spectrum in Tamil Nadu is thus Rs. 315.16 crore. The value of 4.4 MHz of spectrum is computed by multiplying the per megahertz spectrum cost by the number of units of effective spectrum in 4.4 MHz, that is, 2.2 MHz. The value of a 4.4 MHz block is thus Rs. 693 crore. Recall, that the cash flow method gave us a value of Rs. 600 crore.

The values of spectrum using the production function method in metros and category A circles is given in Table 5.7.

In a liberalized environment with operators having the freedom to choose the technology, the value of spectrum will differ for different technologies. Hence, separate valuation exercises need to be carried out for each technology.

Table 5.7 Value of spectrum using the production function method

Circle	Estimated price of 4.4 MHz of 1800 MHz (Rs. crores)
Delhi	451
Mumbai	328
Kolkata	293
Andhra Pradesh	702
Gujarat	593
Karnataka	578
Maharashtra	625
Tamil Nadu including Chennai	693

Source: Authors' own estimates

Conclusion

In a liberalized spectrum environment, allowing any band of spectrum to be used with any technology, regulators may conclude (for instance TRAI 2012) that the relative value of different bands of spectrums is linked only to their relative propagation capability, that is, their 'intrinsic value', thereby ruling out factors related to the 'extrinsic value' of spectrum (Alden 2012; Mölleryd and Markendahl 2012) such as the state of the ecosystem associated with various spectrum bands. As per this reasoning, the 1800 MHz band has 1.2 times the propagation characteristics of the 2100 MHz band and so is 1.2 times more valuable, even though 2100 MHz has a much better developed ecosystem for higher generation technologies. The methods presented stay away from this approach and link the value estimates to the actual performance in specific markets.

The challenge of computing the value of spectrum using the production function methodology is that a spectrum block is not a homogeneous unit. It consists of two different types of spectrums— the first 3 MHz and the rest, with the initial 3 MHz spectrum having 1/3.75 of the capacity, on a per megahertz basis, as compared to each incremental unit. Each spectrum block is rendered homogenous by the transformation of spectrum blocks into equivalent units of 'effective spectrum'.

The merit of the production function approach is that it is able to evaluate the opportunity cost of spectrum at the margin in terms of the cost of physical infrastructure without relying on too many parameters other than production data. However, this approach is less reliable in markets which are very heterogeneous as the estimation of the production parameters may lose statistical validity or the degree of variation explained by the equation estimating the production parameters (the R^2) may become low. Such markets include most category B and category C circles in India which comprise a few urban agglomerations in the midst of far-flung villages with varying population densities.

The cash flow method, on the other hand, relies on a number of operational parameters and is thus onerous from the view of data gathering. However, it may be more reliable in heterogeneous markets.

References

Alden, J. (2012). 'Exploring the value and economic valuation of spectrum', report prepared for the ITU. Available at: http://www.itu.int/ITU-D/ treg/broadband/ITU-BB-Reports_SpectrumValue.pdf. Accessed on 31 October 2012.

Cave, M., C. Doyle, and W. Webb. (2007). *Essentials of Modern Spectrum Management*. UK: Cambridge University Press.

Department of Telecommunications. (2009). *Auction of 3G and BWA spectrum: Revised information memorandum*. Available at: http://www.dot. gov.in/sites/default/files/3g.pdf. Accessed on 10 February 2013.

Miller, Eric. (2008). 'An Assessment of CES and Cobb-Douglas Production Functions', Congressional Budget Office (June).

Mölleryd, B.G., and J. Markendahl. (2012). 'Valuation of spectrum for mobile broadband services—The case of Sweden and India', paper submitted to the regional ITS India Conference 2012, New Delhi, 22–24 February. Available at: http://www.its2012india.com/topics/Spectrum% 20and%20Technology/ValuationofSpectrumforMobileBroadband ServicesTheCaseofSwedenandIndia.pdf. Accessed on 9 November 2012.

Prasad, R. (2010). 'The value of 2G spectrum in India', *Economic and Political Weekly* 45(4): 25–28.

Prasad, R., and V. Sridhar. (2008). 'Optimal number of mobile service providers in India: Trade off between efficiency and competition', *International Journal of Business Data Communications and Networking* 4(3): 69–81.

———. (2009). 'Allocative efficiency of the mobile industry in India and its implications for spectrum policy', *Telecommunications Policy* 33: 521–33.

TRAI. (2010). 'Recommendations on Spectrum Management and Liensing framework.' Available at http://www.trai.gov.in/WriteReadData/ Recommendation/Documents/FINALRECOMENDATIONS.pdf. Accessed on 24 April 2012.

———. (2011). 'Report on the 2010 value of spectrum in the 1800 MHz band.' Available at: http://www.trai.gov.in/Content/Recommendation Description.aspx?RECOMEND_ID=231&qid=0. Accessed on 23 April 2012.

———. (2012). 'Recommendations in response to DOT's reference back on TRAI's recommendations on "Auction of spectrum".' Available at: http://www.trai.gov.in/Content/RecommendationDescription.aspx? RECOMEND_ID=465&qid=0. Accessed on 15 May 2013.

Varian, Hal R. (1992). *Microeconomic Analysis*, Vol. 2. NY: Norton.

6

Spectrum Transitions

Moving from Command and Control to Flexible Use

If we have 3G spectrum everywhere, then we would be investing a lot more in 3G, but we don't have it. I can only invest in the places where we are, which we have already reasonably well covered, maybe invest in some more capacity in some places.
—Marten Pieters, Chief Executive and Managing Director of Vodafone India Ltd., on the 3G roaming case

Introduction

In most countries, the starting point of the liberalization of the telecommunications industry, previously entirely in the public sector, is the command and control approach. As a result of remnants of old ways of thinking, the government initially tries to benefit from the financial resources and operational efficiency of the private sector without giving up control on the objectives and modes of operation, and continues to have a strong public sector presence in the industry. The strong grip that the government has on the industry's operations is significantly based on strict terms and conditions related to spectrum, a critical input in the provision of mobile services.

However, the paradigm of regulation has shifted from pure coordination and planning to the creation of a competitive and sustainable environment for various services, including telecommunications. Technologies have also evolved to accommodate flexibility in spectrum management. As a result countries are migrating in different degrees, and at different speeds to flexible spectrum management regimes.

Command and Control versus Flexible Use

In the command and control approach, the government may do all or some of the following: decide the number of players, select them through an administrative procedure, fix a subsidized price for the licence and spectrum (which is often bundled with the licence), determine specific technologies and services for spectrum use, put in place 'use-it or lose-it' roll-out obligations under which the promoter's stake cannot be sold till time-bound network deployment takes place, tie the assignment of incremental spectrum to achieving subscriber milestones, put in place a usage charge regime that increases with the amount of spectrum held, and impose a universal service obligation.

In contrast, in the flexible use approach, the government uses market mechanisms for spectrum assignment, accords freedom in the choice of technologies and services in those bands, promotes secondary markets for trading and leasing of spectrum, and separates universal service obligations from licence terms. The rationalization of government spectrum is an important element of this approach as it frees up spectrum bands for commercial use. Table 6.1 describes the differences between the two approaches.

The command and control approach is useful for achieving an early roll-out and rapid growth. The low prices of licence and spectrum are believed to keep prices of basic voice telephony low and promote diffusion. The stipulations on spectrum use and roll-out are meant to ensure efficient use of spectrum in the context of spectrum scarcity and the low number of operators. The universal service conditions address the obligation of the government to provide service in underserved regions. However, there are attendant risks of regulatory failures including spectrum getting 'stuck' in lower value uses and absence of mechanisms to put underutilized spectrum to better use.

Table 6.1 Difference between command and control and flexible use regimes

Aspect	Command and Control	Flexible
Allocation of spectrum	Limited number of bands available for commercial use	Greater number of spectrum bands made available, including by rationalizing government spectrum
Spectrum use	Specified technology and service	User-determined technology and service
Assignment of spectrum	Administratively determined price; exclusive use (that is, only primary market in spectrum functions)	Market-determined price (through auctions); trading and leasing of assigned spectrum possible through secondary markets
Roll-out obligations	Mandated	Allows coordination among spectrum owners to meet roll-out targets

Source: Authors' own framework.

The command and control regime begins to fail in the face of several new developments:

(1) Growth of advanced services in new spectrum bands: As 2G technologies give way to 3G technologies that allow advanced services such as VoIP, mobile TV, and video conference, there is no need to subsidize the corresponding spectrum bands. Hence, if the command and control framework were to continue for lower generation spectrums, there will be a dual spectrum regime in which 2G spectrum is subsidized and higher generation spectrums are priced according to the market. The administration of such a regime will become intractable as operators are likely to be holding several bands simultaneously and servicing subscribers using a combination of bands. In such a situation, it may not be possible to segregate subscribers in different bands and apply different charging regimes.

(2) Growth of advanced technologies in old spectrum bands: Spectrum bands originally allocated for voice services may become suitable for advanced technologies. For example, the 900 MHz band originally deployed for 2G technology has developed a rich ecosystem for 3G services. This leads to spectrum getting stuck in lower value uses for the period of the operator's licence.

(3) Convergence of services: The distinction between different types of services, such as landline, mobile, Internet access, and TV broadcasting has blurred. For example, high-speed Internet-access typically provided through landlines is now feasible over wireless connections. Such possibilities make a service-specific licence restrictive.

(4) Growing maturity of competition: In the initial years of the tele-communications industry, there is usually a paucity of spectrum. Hence only a few operators can be introduced. However, as more spectrum gets released and more operators are introduced, competition matures to a point that government strictures are no longer necessary to ensure its efficient use. In India, in 2002 there were 4–6 operators in every LSA. In 2010, the average number of licensees in each LSA had gone up to 15. Any inefficiency in the use of spectrum was sure to be penalized by market forces and did not need to be administratively monitored.

When we move from the command and control approach to the liberalized approach, we have to carefully untangle the different elements of the command and control approach which are bound together by an inherent logic. The task of creating well-functioning spectrum markets lies at the heart of the task. This requires rationalization of the use of government spectrum in order to ensure a good supply of spectrum in the primary market, changing the licensing framework, creating secondary markets for trading and sharing, putting in place a facilitative M&A regime, allowing flexible use of spectrum, and removing universal service obligations from the commercial licence terms.

Rationalization of Government Spectrum

Public sector spectrum in most countries accounts for a major portion of the holding of spectrum below 3 GHz. This spectrum is mainly used for defence purposes. In the UK, for instance, defence accounts for 75 per cent of public sector spectrum. Other uses of public sector spectrum include civil aeronautical, emergency and safety services, and science and maritime applications.

Over the years, the uses of commercial spectrum have grown. Technology innovations in telecom continue to give birth to new applications that interfere with legacy applications. For example, WiMax technology that operates in the 2.5 GHz and 3.5 GHz band, interferes with INSAT communications of the Department of Space (DoS), Government of India. Mobile TV applications that use 700 MHz interfere with traditional TV broadcasting.

As a result, attention has increasingly become focused on whether public sector bodies use their spectrum efficiently. In May 2003, the US government signed a memorandum requiring federal departments to improve the efficiency of spectrum use. In the same year the UK government commissioned an independent audit of spectrum holdings, focusing particularly on public sector spectrum use.

The lack of rationalization in the use of public sector spectrum can lead to a severe paucity of spectrum for commercial use. Table 6.2 indicates the extent of variation in the allocation of spectrum for commercial mobile services across India and Finland and the low availability of commercial spectrum in India.

Public sector spectrum is used for the provision of public goods like national defence and critical communications. Hence the market mechanism is not considered particularly amenable for its allocation.

Table 6.2 Spectrum Allocation for Commercial Services in India and Finland

Factor	India	Finland
Average spectrum allocation per operator per licence service area	2 × 6.5 MHz in 900; 2 × 5 MHz in 1800 for 2G; 2 × 3.6 MHz in 800 for 2G/3G; 2 × 5 MHz in 2100 for 3G; 20 MHz unpaired in 2300 for BWA (Sridhar 2012; Sridhar and Prasad 2011)	2 × 11.3 MHz in 900; 2 × 24.8 MHz in 1800; 2 × 15 MHz in 2100; 4.8 MHz unpaired in 2100; 2 × 20 MHz in 2600 MHz

However, if well-functioning spectrum markets develop that cover a large number of frequencies and provide predictable trading opportunities, then it is conceivable that the public sector can acquire its spectrum on the market rather than by fiat. Conversely, spectrum markets are unlikely to develop unless public sector bodies participate, since the public sector has built up large spectrum holdings in the command and control phase, including in frequencies that are valuable for commercial use.

If users of public sector spectrum are to participate effectively in spectrum markets, certain preconditions have to be fulfilled. Public sector spectrum use is often cloaked in secrecy. In many jurisdictions, government departments are not issued detailed licences specifying rights and responsibilities. Wherever possible, information necessary for potential users to make decisions about the purchase, leasing, or sharing of public sector frequencies has to be made available.

The following intermediate steps can pave the way for bringing more efficiency in public sector use of spectrum and facilitating the growth of spectrum markets overall (Cave et al. 2007). Many of these are currently under consideration by the US government:

(1) Valuation of spectrum: The valuation of spectrum using techniques outlined in Chapter 5 will enable a fuller comprehension of the opportunity cost of spectrum in the public sector.

(2) Audit of actual use: Audit of spectrum use is highly desirable in order to identify areas of underutilization, but may not be feasible due to the sensitive nature of the use of public sector spectrum.

(3) Incorporation of valuation in investment decisions: The value of spectrum should be incorporated as a cost in project appraisals. This will encourage those engaged in procurement to examine the scope of substitution between spectrum and other resources.

(4) Compensation of public sector users: Suppose public sector spectrum has a valuable private use and can be replaced by alternate frequencies, the concerned department should estimate the cost of the transition to the new frequency and be compensated from the auction proceeds of the previously occupied spectrum.

In case the auction proceeds fall short of the cost of migration to the new frequencies, then the process should stop as public sector spectrum has been shown to be efficiently employed in its current use.

(5) Special pre-emption powers for the public sector: Participation of the public sector in the market may be encouraged if there is a clause for the pre-emption of spectrum in an emergency. This clause should not, however, be allowed to undermine the efficiency of the market. Hence public sector users should have to pay market prices for spectrum obtained in this manner; and the procedure should be used sparingly and in strictly defined circumstances.

Responsibility of the Executive Head

Often the National Frequency Allocation Plan (NFAP) that allocates spectrum between different uses and ministries is entrusted to a wireless planning division under the Ministry of Communications, which is also directly responsible for assigning commercial spectrum. This results in the Ministry of Communications having to adjudicate between different peer ministries (for example, defence and information and broadcasting) for the allocation of spectrum. This adjudication is often stalled because of the conflict of interest involved and the absence of a suitable adjudicating authority. As a result, inter-ministerial disputes are not handled with an appropriate process causing unnecessary delays and inconsistencies in the allocation of spectrum.

Ideally, spectrum allocation should fall under the purview of the office of the chief executive of the country which has the necessary standing to settle the allocation of spectrum between ministries and chart out a strategy for achieving the country's strategic goals in international forums like the ITU. Countries such as France have recognized the importance of spectrum as a key national resource and brought it under the direct supervision of the executive head.

A useful case study in the reallocation of spectrum from government to commercial use lies in the release of 'digital dividend spectrum' in India.

Release of Digital Dividend Spectrum in India

The digitalization of terrestrial TV networks has started in most of the advanced markets. This process frees-up a significant amount of the ultra high frequency (UHF) band which can be potentially allotted for commercial mobile services. This spectrum, referred to as the 'Digital Dividend' spectrum, varies between countries and regions, but amounts to approximately 100 MHz (108 MHz in the US, 128 MHz in the UK, 72 MHz in France, and 54 MHz in Korea). Since the Digital Dividend spectrum is in the UHF range, it has very good propagation characteristics (being less attenuated by obstacles such as buildings). It is approximately 70 per cent cheaper to provide mobile broadband coverage over a given geographic area using UHF spectrum than with the 2100 MHz spectrum widely used for mobile broadband today. The low-frequency characteristic makes this spectrum particularly well-suited for providing mobile broadband coverage in rural and suburban areas.

In June 2009, all the US terrestrial broadcasting stations switched off their analog transmissions and turned on digital transmissions, thus freeing up spectrum for mobile and emergency services. The European Commission issued a mandate in 2008 to the Conference of European Post & Telecommunications (CEPT) to carry out investigations to define technical conditions for the use of the 790–862 MHz Digital Dividend spectrum by fixed/mobile communication networks (Karimi et al. 2010). The switchover from analogue to digital terrestrial television has been completed in many of the EU member countries and is expected to be completed in the whole of Europe by the end of 2015. Finland started the TV digitalization process in June 2008. A spectrum block of 2 × 30 MHz in the 791–862 MHz block has been made available for commercial mobile services.

In India, the relevance of over-the-air broadcasting has reduced over the years due to cable and direct to home (DTH) satellite television. Currently only the government operator provides over the air broadcasting. There are about 60,000 local cable operators and 7 DTH providers in the country with close to 80 per cent of the 135 million TV sets either connected to cable TV or DTH. Further the government has set a deadline of digitalization of cable TV. The first phase of cable TV digitization has been relatively successful. If the second phase proceeds smoothly, India will be well on its way to joining the elite club of many advanced countries in sun-setting analogue television. However, though the Telecom Regulatory Authority of India (TRAI) specified 585–862 MHz for mobile TV services way back in 2008 (TRAI 2008), the spectrum has not yet been released due to continued tenancy by the Ministry of Information and Broadcasting. More details on the Digital Dividend are discussed in Chapter 11.

New Licensing Framework

In most countries, service and spectrum licences are delinked from each other. The service licence is obtained by paying a nominal annual fee and is automatically renewed in the normal course. It includes standard terms on doing business and may relate to a set of services (or may be a universal service licence allowing all telecommunications services). Spectrum licence is allocated to licensees for fixed periods of time based on certain methods including auction and may specify a technology (for instance, 2G) or may be open to flexible use.

In those countries where the service licence comes bundled with spectrum, explicit delinking of licence from spectrum is required to move to a market-oriented mechanism for spectrum. An unbundled spectrum licence is amenable to trading and leasing, including to parties that may not hold an operator licence. When the service licence and spectrum licence are bundled together, a mismatch between the remaining licence tenure of the buyer and seller can exert a dampening effect on transactions. For instance, if the remaining licence tenure of the seller is greater than that of the potential buyer, the seller may need to find a buyer for the excess tenure of the licence on sale. The necessity to remove clauses related to spectrum use from the spectrum licence is detailed in the section 'Flexible Use'.

Annual Spectrum Usage Charge

One of the features of the licensing framework of command and control regime is an annual spectrum charge that varies with the amount of spectrum held. The escalating percentage is a way of charging for additional spectrum which does not attract an upfront payment.

It is also meant to bring about the efficient use of spectrum. With a market-oriented mechanism such as an auction, and sufficient competition in the market, escalating annual spectrum charges are no longer required. Such charges would create unnecessary heterogeneity between bidders holding different quantities of spectrum at the time of bidding. This will also create complications in the trading and leasing of spectrum (described in the section 'Creating Secondary Markets').

The government may desire that migration to uniform rates of usage charges should be revenue-neutral. In view of the additional revenues accruing from auctions, the government need not worry unduly about the loss of revenue on account of annual usage charge. Sridhar (2006) argues that regulatory levies such as spectrum charges should be in proportion to the cost of administration. However, if there are political compulsions for maintaining revenues from usage charges at previous levels, then one option would be to take the average usage charge rate for the industry and fix this as a uniform rate. Uniformity of annual spectrum charges across all frequency bands and technologies will facilitate change of use and effective use of assigned spectrum.

Creating Secondary Markets

Secondary markets in spectrum allow its trading and leasing. The presence of such markets reduces the risk for participants in primary markets.

Trading of Spectrum

Crocioni (2009) defines trading as a process in which there is a change in ownership of a licence for economic considerations. In a trading regime, licensees are allowed to aggregate and disaggregate their spectrum endowments (either by geography, frequency, or time). Thus, spectrum gravitates to those who value it most, allowing for the establishment of dynamic and competitive wireless communication markets. Caicedo and Weiss (2011) as well as Crocioni (2009) point out that such spectrum markets can be viable if sufficient numbers of market participants exist and the amount of tradable spectrum is balanced to the demand. Trading can be via mutual agreements between the parties or via exchanges as indicated in Caicedo and

Weiss (2011). A uniform spectrum usage charge homogenizes the taxes faced by the buyer and the seller and promotes transactions.

In the active secondary spectrum markets in the US, mobile operators have bought spectrum from each other as well as from broadcasters and other niche spectrum holders. Apart from acquiring spectrum directly, another option is of acquiring the spectrum holding firm. Of late, AT&T and Verizon have been on a buying spree as they vie with each other in expanding their 4G LTE networks. Since the 2011 collapse of AT&T's $39 billion bid for T-Mobile USA, it has been working to acquire spectrum from a variety of other sources. For example, in 2012 AT&T obtained approval to purchase 700 MHz and 2300 MHz spectrum from the likes of NextWave Wireless, Comcast, Horizon WiCom, and the San Diego Gas & Electric Company. In 2011, AT&T also bought a sizable chunk of spectrum from Qualcomm, which the chip company had used for its failed 'Flo TV' venture. Verizon recently agreed to pay $3.6 billion to buy spectrum from a consortium of cable companies to augment its spectrum capacity. AT&T completed the acquisition of Leap Wireless in March 2014, and will be deploying 4C-LTE services in the spectrum band currently used by Leap. Table 6.3 indicates cases of trading among different operators.

Table 6.3   Examples of partnerships in spectrum trading

Year	Transaction	Valuation
2011	Verizon Wireless bought 700 MHz spectrum from Spectrum Co, a group including Time Warne Cable, Comcast Corporation, and Bright House Networks	$3.6 billion
2011	AT&T bought 700 MHz from Qualcomm	
2013	AT&T is planning to buy 700 MHz spectrum from Verizon Wireless	$1.9 billion
2013	Verizon planning to buy 2.5 GHz spectrum from Clearwire	$1.5 billion
2014	AT&T acquiring Leap Wireless for augmenting its spectrum holding in Advanced Wireless and PCS bands to offer 4G-LTE Services	$1.2 billion

Source: Data collected from various sources by the authors.

As can be seen from Table 6.3, most of the trading is in 700 MHz due to the broader coverage of this spectrum and the accompanying ecosystem for 4G LTE services.

Though spectrum trading studies were initiated in Europe around the turn of the millennium, it is only recently that country regulators have given the go-ahead for spectrum trading between MNOs. OfCom, the telecom regulator in the UK, allowed spectrum trading in 900 MHz, 1800 MHz, and 2100 MHz in 2011 which was followed by the recent announcement on 5 April 2013 for allowing trading in 800 MHz and 2600 MHz. However, not much action has taken place as yet in the European market. Spectrum trading is not yet allowed in India.

Enabling Sale of Inefficient Spectrum

Potential sellers of spectrum include those who are unable to make optimal use of it. In the command and control regime, such operators may have got spectrum cheap, and stand to make windfall gains by selling in spectrum markets. In many cases such a gain comes with a political risk for the government which may be seen as having transferred national wealth into private hands. To guard against this, the government introduces a lock-in period for stake sales by a promoter, that is, someone with significant share capital in the licensee, and whose net worth has been taken into consideration when determining eligibility for granting the licence. If the firm fulfils all the network roll-out obligations, then the lock-in period is waived. The main objective of such rules is to block unearned gains arising from transactions in promoters' stakes particularly when the value of spectrum is not getting correctly reflected in the licence fee.

Without the participation of such licensees, the sellers' side of the secondary market in spectrum could be virtually non-existent as it is quite likely that incumbent licensees may be unwilling to part with any of their spectrum. This creates market 'thinness' as explained by Bykowski (2003). Therefore the lock-in period should be waived provided the consequent opportunity for unearned windfall gains is addressed. Such windfall gains can be moderated by levying a spectrum transfer charge on all such transactions. The case for levying a tax on net windfall gains from spectrum trades should be balanced by the consideration of encouraging the efficient use of spectrum.

The importance of having low transaction costs in ensuring an efficient market in spectrum has long been recognized (Coase 1960). The spectrum transfer charge should be chosen such that it provides adequate rate of return for both the buyer and the seller of the transaction. While such an approach to fixing the rate of transfer charge may not ensure that the full value of the spectrum accrues as revenue to the government, it will ensure that the spectrum reaches the hands of an entity that values it the most and puts this scarce resource to its most efficient and optimal use. It is this efficient and optimal use of the spectrum resource that should be the primary objective of the government, rather than the maximization of revenues. In other words, the trading policy should create a conducive environment for spectrum sale without undue windfall gains to licensees.

Leasing of Spectrum: Spectrum Manager Model

On occasion, operators may want to enter into agreements to lease spectrum without transferring the rights and obligations of the original licence holder (Sridhar and Prasad 2011) who continues to act as a spectrum manager. This is sometimes referred to as spectrum sharing although that term is more appropriate for situations where two entities use a common block of spectrum in a non-exclusive manner. In this book we use 'sharing' to refer to only *non-exclusive* use of a spectrum block by two or more operators.

Leasing can be beneficial (a) among operators having non-uniform and complementary subscriber base in different parts (say, urban and rural) of a service area, that is, intra-region leasing, and (b) when operators hold spectrum in complementary service areas and want to provide national coverage to their subscribers, that is, inter-region leasing. While parties in a leasing agreement obtain exclusive rights to certain frequency bands, the original licence holder continues to be responsible for abiding by the licence's terms and conditions. In case the original licence holder has acquired the spectrum at subsidized rates and is within the lock-in period, leasing charges should be levied in a manner similar to transfer charges in the case of trading.

Intra- and Inter-region Leasing

Instances where the need for leasing exists are commonplace. In India, the 3G and Broadband Wireless Access (BWA) auctions took place in 2010. 2 × 5 MHz paired frequency blocks in 2100 MHz and 20 MHz unpaired frequency blocks in 2300 MHz for 3G and BWA, respectively, were auctioned. Bids for pan-India spectrum were as high as $3.7 billion and $2.85 billion, respectively. Inadequate spectrum and bullish expectations on wireless broadband demand were possible reasons for the high bids. However, due to escalating prices, none of the bidders got 3G spectrum for all the 22 LSAs. In the BWA auction held after the 3G auction, all bidders except one received spectrum for only 4–5 LSAs. The disparity in spectrum allocation across different LSAs for wireless broadband services (see Table 6.4) creates the need for inter-region spectrum leasing to provide services across India's entire geography.

In Finland, 3G spectrum was allotted using administrative hearings where applicants presented their financial and technical credentials for acquiring spectrum ('beauty contests'). Though all operators have been given national licences, not all of them have 100 per cent of the mainland area covered. Even the Finnish archipelago needs coverage due to habitation during the summer season. Since the demand for network access in these areas is seasonal and often not bandwidth-intensive, there is a case for leasing spectrum and the associated infrastructure (for example, cell sites and backhaul capacity) to minimize cost (Sridhar et al. 2013).

An operator who does not have radio access infrastructure in specific geographical areas can possibly use the spectrum and the associated infrastructure of other existing operators to provide coverage. This is often referred to as 'national roaming'. National roaming accelerates competition by allowing new operators to provide services within shorter time frames. Though national roaming agreements normally have a sunset clause, they can be made mandatory in specific locations, especially in rural and remote areas (Shanab et al. 2007). Though leasing of spectrum is not allowed in India, 3G operators have already started this practice in select areas in the country.

Administering leased spectrum is difficult when spectrum usage charges are non-uniform across different blocks of spectrum.

Table 6.4 Disparity of spectrum allocation for wireless broadband services in India

LSA	Op-1	Op-2	Op-3	Op-4	Op-5	Op-6	Op-7	Op-8	Op-9	Op-10	Op-11	Op-12	Total
1	3G+BWA	BWA	3G	3G		3G				BWA			6
2	3G+BWA	BWA	BWA	3G		3G		3G					6
3	3G+BWA	BWA	3G	3G		3G				BWA			6
4	3G+BWA	BWA	3G		3G			3G+BWA					5
5	3G+BWA	BWA			3G	3G	3G				BWA		6
6	3G+BWA	BWA	3G+BWA			3G	3G						5
7	3G+BWA	BWA	BWA		3G	3G	3G						6
8	3G+BWA	BWA	3G			3G		3G+BWA					5
9	3G+BWA	BWA			3G	3G			BWA	BWA			6
10	3G+BWA	BWA			3G			3G	BWA	BWA			6
11	3G+BWA	BWA		3G	3G	3G						BWA	6
12	3G+BWA	BWA	BWA	3G	3G	3G	3G						7

#										Total
13	3G+BWA	BWA	3G	3G			3G			BWA	6
14	3G+BWA	BWA		3G	3G		3G			BWA	6
15	3G+BWA	BWA	3G	3G	3G	3G				BWA	6
16	3G+BWA	BWA	3G	3G		3G		3G+BWA			6
17	3G+BWA	BWA	3G					3G+BWA			5
18	3G+BWA	BWA	3G	3G				3G+BWA	3G		6
19	3G+BWA	BWA	3G	3G					3G	BWA	7
20	3G+BWA	BWA	3G	3G				3G+BWA	3G		6
21	3G+BWA	BWA	3G					3G+BWA			5
22	3G+BWA	BWA	3G					3G+BWA	3G		5

Source: Sridhar et al. (2013).

The original licensee will be liable to pay usage charges commensurate to the entire spectrum block, even though a smaller block of spectrum is directly controlled on which a lower usage charge may be payable. On the other hand, the lessee may get away by paying a rate commensurate with the size of the block he controls. Setting a uniform usage charge will help avoid these issues.

Leasing between MNOs and MVNOs/ISPs

Economical and technical efficiency gains, especially in markets such as India, can also accrue when there are pockets in an LSA where an operator has underused spectrum and wants to lease it to Mobile Virtual Network Operators (MVNOs),[1] Internet service providers (ISPs), venue owners (that is, hotels, hospitals, malls, airports), or Femtocell operators.

It is to be noted that such transactions do not involve a change of property rights. The original licensee continues to be responsible for abiding by the licence's terms. In some cases, few obligations also devolve directly upon the MVNO. The possibility of such trades will increase the demand for spectrum in markets where property rights are exchanged.

Mergers and Acquisitions

In some cases, value can best be created not merely by trading or leasing spectrum but by and M&A transaction. In case such a transaction takes place during the lock-in period, similar considerations as in the case of trading and sharing of spectrum will apply. Hence, M&A should only be permitted on payment of 'merger charges' to the government for the quantity of spectrum shared, in the same manner and of like amount as applicable in case of transfer of the spectrum.

The competition regulator of the telecommunications sector often sets an upper bound on both the market share of the combined entity as well as the percentage of spectrum held. For instance, the M&A

---

[1] Mobile Virtual Network Operator is a licensed/registered telecom operator who normally leases/rents spectrum and other associated facilities to provide mobile communication services.

regime in India currently allows the merged entity to have up to 35 per cent of the market share, with up to 60 per cent being allowed on a case-by-case basis; but only 25 per cent of the total spectrum. It is necessary to set two bounds because an entity with a low market share but high spectrum share can exercise significant market power in the future.

Sometimes, as is the case of India, regulators set a high cap on market share but a low cap on spectrum share. The criteria chosen by the regulator are perhaps intended to reward companies which utilize spectrum efficiently and thereby have a large share of subscribers with a low share of spectrum. However, in a densely populated market, it is technically not feasible to have a high percentage of subscribers with a low percentage of the allotted spectrum. This may be possible in less dense circles. But such circles are 'coverage-constrained' and not 'capacity-constrained', that is, the bottleneck is physical infrastructure not spectrum. In such circles, there is a case for allocating some of the licensed spectrum for the commons after which there will be parity between the share of licensed spectrum and market share.

Many viable mergers bring together a company that utilizes spectrum efficiently and one that does not utilize spectrum efficiently but possesses a sizable amount of it. The recent merger of AT&T, the efficient operator, and T-Mobile, the inefficient operator, with a large holding of spectrum is a case in point (FCC 2011). Mergers of such companies will enable spectrum to be utilized more efficiently. But such recommendations, in many cases, prohibit such transactions. Many of the transactions that are allowed are unlikely to occur as they bring together super-efficient operators who would rather compete with each other.

Femtocells

Femtocells are small micro/pico cells that are deployed as in-building solutions operating in licensed bands. Femtocells allow traffic to be diverted from a carrier's macro cellular network to a localized network, thus relieving the load on the macro network's licensed spectrum. The released spectrum can be re-used for other conversations. Femtos use low power access points deployed inside buildings and communicate with the public switched telephone or data networks over a

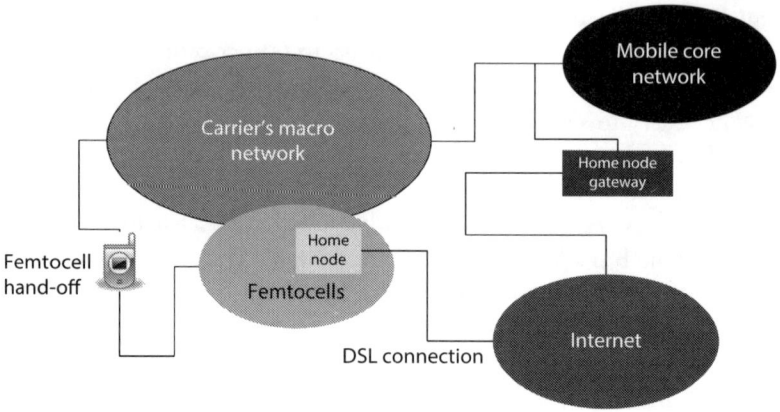

Figure 6.1   Femtocell architecture

broadband connection such as a digital subscriber loop (DSL) or a cable modem. Standard mobile cellular network protocols, such as GSM, CDMA, WCDMA, LTE, or Mobile WiMax, are used to communicate with the mobile handset. Though Wi-Fi at home resembles Femtocell, it requires dual-mode handsets due to its operation in unlicensed spectrum bands. On the other hand, Femtos can deliver both in-home and mobile voice and data services with existing handsets. They require a certain portion of spectrum holding to be re-farmed by the operators.

The architecture of a typical 3G Femtocell is given in Figure 6.1.

While the macro radio network connects the subscriber to the mobile core network outside the house, the hand-off to the home node takes place once the subscriber gets in the house. From thereon, the subscriber generated voice and data traffic is captured by the home node and sent through the broadband connection over the Internet to the home node gateway (HNG), which then passes it on through defined interfaces to the mobile core network. It is to be noted that the home node incorporates both the functions of BTS as well as radio resource management present in the Radio Network Controller (RNC). This is referred to as flat architecture. HNG installed in the operator's network is a very important component in the Femtocell network architecture. The HNG acts as a concentrator aggregating traffic from a number of similar home nodes, thus obviating the need for the operators' core network to deal with millions of home nodes. This enables the home nodes to be deployed seamlessly without any modifications to the operators' core network functions.

Femtocell home base stations deployed in combination with the macro-cellular network have the potential to significantly reduce total network costs. The fundamental reason for the savings is that the macro network radio frequency coverage is overburdened when attempting to provide indoor coverage, especially due to severe signal attenuation inside homes. The operator incurs significant capital and operational expenditure to provide more macro coverage. Also, studies have proven that while the savings in costs for the operator are more for high bandwidth data traffic, operational efficiencies accrue even in the case of traditional voice services. Moreover, the Femtocell box sitting in a user's home can learn the usage pattern and hence can provide valuable data to the operators to provide personalized service offerings. While Femto Forum (www.femtoforum.org) with over 100 members (including fixed, mobile, and integrated operators) is actively promoting the use of Femtocells, commercial deployment is still limited to certain pockets (Denver and Indianapolis in the US) and the first set of launches in Europe and Japan are expected only this year. Femtocells could prove to be a disruptive technology that provides a win-win for operators as well as subscribers. The cost of a Femtocell is as low as $50 (the cost of a TV set-top box; less than a Wi-Fi access point) and is slated to go down further as volumes pick up and hardware technology improves.

Since Femtocells can coexist in the same spectrum as macro cells, there is no need for specific femtocell spectral allocations. Although initial deployments have used separate or partially separate bands for deploying Femtocells there is significant pressure on operators to move to shared carrier deployments. Operators also need an approach that seamlessly works across countries and regions, minimizing configuration and special settings, thus minimizing operational costs. Nevertheless, there has been some interest in Femtocell-specific allocations. For example, the UK regulator OfCom has proposed to allocate a portion (as much as 2 × 20 MHz) of the 2.6 GHz band specifically for low-power use (Andrews et al. 2011). Even in cases, where niche Femtocell operators exist, they use the spectrum of their associated mobile operators in the Femtocells.

Table 6.5   Types of transactions in spectrum in secondary markets

	Licensed Spectrum			
	Spectrum rights not transferred (leasing)		Spectrum rights transferred	
	Between operators	Between operators and other entities	Between operators	Between operators and other entities
Exclusive use	Intra-circle and inter-circle roaming	MVNO, Femtocells	Trading, mergers and acquisitions	NA

Source: Authors' own framework.

The various kinds of transactions possible in secondary markets are summarized in Table 6.5.

Flexible Use

Procedures for changing use or other parameters of the licence vary significantly across countries. For example, in Australia and New Zealand, property rights are broadly defined according to technical or core parameters that set the maximum level of emissions. If this level is exceeded, the affected licensee has a right to force the licensee which causes interference to take measures to reduce its emissions. However, the licensee is free to act as it wishes including modifying the type and nature of the services offered and the technology used, as long as interference guidelines are followed (Crocioni 2009).

In many countries in the early stages of spectrum liberalization, there are rigid guidelines on services and technologies. For instance, in India, spectrum assigned for 2G use in the 800 MHz and 1800 MHz bands, cannot be used to provide 3G/4G services even though there are vibrant ecosystems for the provision of CDMA-EVDO Rev B, and LTE based 3G/4G services on these bands. 3G services can be provided only in the 2100 MHz band. The use restrictions hinder the value of spectrum in the hands of the buyer and hence the incentive to trade. Further, existing commercial users of spectrum have very little incentive to sell excess or unused spectrum if the buyer will use its acquired spectrum to provide a service that is currently provided by the seller (for example, 2G or 3G).

Consequently, the number of participants in such a spectrum market is likely to be very low. Such market 'thinness' decreases the likelihood that a trade will take place. For example, in Australia the low activity levels in the trading market are attributed to lack of portability and the nature of property rights specified in the licence (Xavier and Ypsilanti 2006). Hence, change of use of a spectrum band should be allowed after trading so that the buyer deploys the acquired spectrum block in its most effective use. Such a spectrum trading market will allow firms to choose technologies based on market conditions rather than on standards dictated by the government. This will also allow a de facto standard to emerge through competition, if appropriate.

Refarming

In a command and control regime, spectrum bands allocated to a lower generation technology are migrated to higher generation technologies in the event of suitable technological and commercial developments. For instance, the 900 MHz GSM band is being migrated to 3G/4G technologies in India. The band being migrated is held by two kinds of operators—those whose licences are coming to a close and those who still have significant time remaining on their licences. A common practice in the case of operators whose licence periods are coming to a close (also followed in India) is to take back their spectrum for 'refarming' and allocating an equal amount of spectrum suitable for lower technologies (for example, 2G) at a market determined price.

This practice draws flak from operators on a number of counts. First, they contend that the transition to the new spectrum is costly. Refarming makes the equipment used in previously held frequencies partially unusable. Hence getting an equal number of units in lieu of the spectrum they held is not a fair deal. Second, the replacement spectrum may have lower propagation characteristics, as is the case in the current episode of refarming in India from 900 MHz to 1800 MHz. This increases the cost of the physical infrastructure.

The operators' objections may not be entirely valid. The ostensible objective of the government in refarming is to ensure continuity of business of the incumbent by guaranteeing a certain amount of spectrum at the market determined price while at the same time migrating spectrum bands to modern technologies. However, there is no obligation on the government to assure incumbents an equivalent spectral capacity at the time of licence renewal. Indeed, even the limited measures towards continuity, while common in spectrum jurisdictions, may be regarded as unnecessary, especially when a number of spectrum bands are available in auctions. The onus of retaining enough spectrum to provide uninterrupted services to subscribers rests with the operator, not the government. The welfare of the subscriber in turn will be ensured by market competition, and the USO fund. The government's responsibility is the continuation of the service and not that of a specific operator.

Reserving a certain amount of lower generation technology spectrum for operators renewing licences reduces the amount of spectrum that is auctioned, and thereby distorts the determination of market price. This distortion is further accentuated by the fact that the cost of licence renewals depends on the price discovered in the auctions, thereby dampening the bids of affected operators. A more suitable approach would be to take back all spectrum from those whose licences are expiring and invite them to bid in the auction to acquire the frequencies they need. In a liberalized spectrum regime, these frequencies can be used in any manner by the winning bidders subject to limits on the power of emissions, as in the case of New Zealand and Australia. Alternatively, in a command and control regime, they can be used for specific technologies stipulated by the government. In case the operator does not want to participate in the auction or loses in the bidding, he has the option of bidding for other spectrum bands in order to service his customers.

The second category of operators, those who have some years remaining in their licence, should be allowed to change use provided they pay the market determined rates for the spectrum that they hold. The argument for continuity has relevance in the case of in-term licences. Taking back spectrum from such licensees would disturb the stability of the licensing regime.

Operators often argue against refarming practices because they want to retain control over valuable spectral resources on the pretext of continuity of business. The cost of migration to new spectrum bands is lower than the cost of setting up a new network. Hence in an auction for this particular spectrum band where new operators and incumbents participate, this cost will be factored into the bids, and will be reflected in the market price. In order to maintain their network capacity they can bid for more units of spectrum or choose another spectral band entirely, in line with their business plans.

Overall, refarming continues the old paradigm of command and control in which the government decides the technology to be used in a particular band. Moving to a truly flexible use regime would make refarming irrelevant. This is the goal that regulators should strive to attain.

Smaller and More Frequent Auctions

Researchers have pointed out that when small blocks of frequencies become available at different intervals of time as in the case of India, there may be benefits from holding smaller and more frequent auctions as soon as spectrum becomes available rather than waiting and assigning the whole spectrum in one go. For example, the number of auctions held since 2002 has been particularly high in the US with 29 instances and fewer in Australia (4), New Zealand (4), and Norway (5). In the UK, a handful of auctions have been held after spectrum trading was introduced. In India also, spectrum should be auctioned in small blocks of 2 MHz x 1 MHz, as and when available. This procedure will provide enough quality information to the secondary market as well as regarding the value of the spectrum, thus making it efficient.

Universal Service Obligation

In the C&C paradigm, the universal service obligation is written in the licence terms of the operators in the form of certain obligations for rural roll-outs. However, this creates a conflict between an operator's profit optimizing, growth objectives, and obligations towards the public policy goal of providing universal service. It is best to decouple the two objectives. In the liberalized paradigm, a percentage of operators' revenues is apportioned for providing universal connectivity and the funds can be administered by a universal service operator under the government. The percentage apportioned varies from 0.5 to 5 per cent. The following points need to be remembered in designing the universal service scheme:

(1) The percentage of revenues apportioned should be calibrated to the need and ability of the universal service organization to spend. An Infodev study of 15 USO funds in developing markets showed that only 26 per cent of the US$ 7 billion had been disbursed by 2006.
(2) The projects should be reverse-auctioned to the best bidder and there should be no preference for the public operator.
(3) As the uptake of ICT requires inter alia literacy, digital literacy, and relevant applications, the administrator should have some ability to coordinate the projects across ministries in order to ensure that the projects under the USO have the maximum impact. One possibility is creating a national broadband mission under the executive head.

Conclusion

This chapter dealt with individual elements of the command and control regime and their transition to a framework of flexible use. Usually countries adopt a gradual approach to the liberalization of spectrum. In India, spectrum auctions and leasing have been introduced, but trading appears to have been ruled out for now. The government continues to determine the use and technology to be deployed, and undertakes refarming to upgrade to advanced technologies. The release of the Digital Dividend spectrum has been cleared by the regulator but the rationalization of government spectrum is not on the cards on account of its politically sensitive nature. This chapter provided a set of practical tips that governments can use to tailor their individual paths to flexible spectrum use. The types of partnerships that enable a flexible spectrum regime and specifics on the India case are dealt with in later chapters.

References

Andrews, J., H. Claussen, M. Dohler, S. Rangan, and M. Reed. (2011). 'Femtocells: Past, present, and future', *IEEE Journal on Selected Areas in Communications* 30(3): 497–508.

Bykowski, M. (2003). 'A secondary market for the trading of spectrum: Promoting market liquidity', *Telecommunications Policy* 27: 533–41.

Caicedo, C.E., and M.B. Weiss. (2011). 'The viability of spectrum trading markets', *IEEE Communications Magazine* 49(3): 46–52.

Cave, M., C. Doyle, and W. Webb. (2007). *Essentials of Modern Spectrum Management*. UK: Cambridge University Press.

Coase, R. (1960). 'The problem of social cost', *Journal of Law and Economics* 3: 1–44.

Crocioni, P. (2009). 'Is allowing trading enough? Making secondary market in spectrum work', *Telecommunications Policy* 33: 451–68.

Federal Communications Commission (FCC). (2011a). *AT&T and T-Mobile—WT Docket No. 11–65*. Available at: http://transition.fcc.gov/transaction/att-tmobile.html. Accessed on 15 May 2012.

Karimi, H.R.F., M. Lapierre, and E.G. Fournier. (2010). 'European harmonized technical conditions and band plans for broadband wireless access in the 790–862MHz digital dividend spectrum', *2010 IEEE Symposium on New Frontiers in Dynamic Spectrum*, 1–9, Singapore: IEEE, 6–9 April.

Shanab, L.A., B. El-Darwiche, G. Hasbani, and M. Mourad. (2007). *Telecom Infrastructure Sharing: Regulatory Enablers and Economic Benefits*. Available at: <http://www.booz.com>. Accessed on 20 March 2013.

Sridhar, V. (2006b). 'How much do you pay as telecom taxes?' *Financial Express*, 25 August.

———. (2012). *Telecom Revolution in India: Technology, Regulation and Policy*. New Delhi, India: Oxford University Press.

Sridhar, V., and R. Prasad. (2011). 'Towards a new policy framework for spectrum management in India', *Telecommunications Policy* 35(2): 172–84.

Sridhar, V., T. Casey, and H. Hämmäinen. (2013). 'Flexible spectrum management for mobile broadband services: How does it vary across advanced and emerging markets?' *Telecommunications Policy Special Issue on Cognitive Radio* 37: 178–91.

TRAI. (2008a). 'Recommendations on issues relating to mobile television service.' Available at: http://www.trai.gov.in/WriteReadData/Recommendation/Documents/FINALRECOMENDATIONS.pdf. Accessed on 23 April 2008.

Xavier, P., and D. Ypsilanti. (2006). Policy issues in spectrum trading. *Info* 8(2): 34–61, doi:10.1108/14636690610653581.

7

Pathways to the Spectrum Commons

Today, unlicensed wireless devices contribute between $16 billion and $37 billion to our economy annually. To put that in perspective, that is more than Americans spend on milk and bread each year, combined.

— The U.S. Federal Communications Commission's
Commissioner Jessica Rosenworcel on Growing
Unlicensed Spectrum and Unlicensed Economy

Introduction

New technologies mitigating the risk of interference are creating the possibility of new frameworks of spectrum management. In Chapter 2, we highlighted the conceptual aspects of the debate between privatization and the commons. In this chapter, we present a historical perspective on different models for the sharing of spectrum, and highlight emerging trends in the non-exclusive use of spectrum using both licensed and unlicensed spectrum bands.

History

The spectrum commons has a long and hoary history. It was the original form of spectrum management—in the 1920s, essentially all spectrum was unlicensed. Anyone possessing radio equipment was entitled to broadcast signals over the air. The resulting interference caused radio sales to plummet.

This led to the passing of the Communications Act of 1934 under which the Federal Communications Commission (FCC) was set up to regulate radio communications within the United States. The FCC has historically controlled access to radio spectrum by allocating specific frequency bands for use by licensed service providers. The use of devices on unlicensed spectrum was first authorized by the FCC in 1938. At that point devices were approved on a case-by-case basis. The initial qualifying devices included wireless record players and remote control devices. Over time, provisions were made to permit the operation of wireless microphones, garage door openers, telemetry systems, security alarms, and cordless telephones. The choice of devices was by and large based on the low power level of the emissions that they were required to make in order to be effective in the limited area in which they were expected to be functional.

Advances in technology made it possible to include devices that needed to function over larger areas. In 1985 the FCC eliminated the process of case-by-case approval and set forth technical criteria which the new unlicensed devices had to adhere. The FCC also opened up new spectrum for unlicensed use at 902 MHz–928 MHz, 2400 MHz–2483.5 MHz, and 5725 MHz–5850 MHz. The release of new bands facilitated the development of Bluetooth and Wi-Fi devices that have become ubiquitous. These technologies use spread spectrum techniques, originally developed by the military, which provide high immunity to interference noise as compared to conventional techniques and allow more devices to operate in a given frequency band, thus promoting more efficient spectrum use.

Over the last 25 years, the FCC has made further bands available for unlicensed use. The 59 GHz–64 GHz band, for instance, which facilitates high bandwidth wireless communications between electronic devices over short distances, was made available in 1995. In the US as of the end of 2008, approximately 955 MHz were allocated to unlicensed uses below 6 GHz.

In the past 10 years, interest in greater use of unlicensed spectrum has grown sharply. Deployments of new technologies in the 2.4 GHz band, particularly WLANs, have been commercially successful. Some of the standards and products in the unlicensed bands are:

WLAN and Wi-Fi: A WLAN is a network that connects two or more electronic devices in a limited area using radio spectrum. Users enjoy the ability to move around in the coverage area while being connected to the network.

WPANs (Wireless Personal Area Networks) and Bluetooth: WPANs are computer networks designed for communication between electronic devices within close proximity of each other. The most widely adopted form of WPAN is Bluetooth, a technology standard that is implemented in a wide variety of applications such as mobile phone headsets, PC peripherals such as mouse, keyboards, and printers.

Wireless HD and WiGig: Wireless HD and WiGig use the 60 GHz unlicensed band to achieve multi-gigabit data transfer over the range of a few metres. Applications of these technologies include home entertainment, data networking, and wireless docking.

RFID (Radio Frequency Identification): RFID is the use of a small chip or tag embedded in cards and attached to products, animals, or vehicles for purposes of identification or tracking. Applications include supply chain management, asset tracking, sports event timing, medical applications, and payments for toll and transit fares.

DLNA (Digital Living Network Alliance): DLNA is a set of protocols advocated by the DLNA group to connect various devices at home including wireless speakers, televisions, set top boxes, mobiles, tablets and PCs over home Wi-Fi networks to share images, music files, and videos.

Vehicular networks: These are in-car infotainment systems that connect mobile devices with a rear seat and front seat video and audio system using Wi-Fi and Bluetooth and are being promoted by car connectivity consortiums. The US National Transportation Safety Board is about to mandate that motor vehicles be equipped with 'connected technology' that can help drivers avoid accidents. Currently, researchers are developing a machine to machine (M2M) communication technology that will enable vehicles to exchange data and know what is taking place around them wirelessly.

Super Wi-Fi: Super Wi-Fi uses lower frequency spectrum to increase the range normally provided by Wi-Fi by two or three times, as well for providing connectivity over hills and through walls. This is especially suited for rural areas. The technology uses white spaces—the unused guard bands of spectrum between adjacent channels, in the upper 700 MHz band. In order to prevent conflicts and interference, new standards such as IEEE 802.22 incorporate technologies such as spectrum sensing, dynamic spectrum access, and geo-locational techniques.

Much of the debate regarding further allocation of unlicensed spectrum focuses on frequencies below 1000 MHz as these are frequencies that are of most use to the traditional narrow band, high power, long distance transmitters. Radio waves at these frequencies travel further and penetrate walls more easily for a given power level than those at higher frequencies. These characteristics make many of these frequencies especially useful in both licensed and unlicensed applications, from 4G mobile services to long range Wi-Fi.

Municipal wireless networks are a potential application that could emerge with additional unlicensed spectrum in the lower bands. An example of such a network is the Paris Wi-Fi, which provides free Internet access in many parks and municipal libraries, museums, and other public places. Currently municipal Wi-Fi networks require a dense installation of transmitters due to the low permissible power levels in unlicensed bands. This requirement can be eased if Wi-Fi can use the lower end of the spectrum.

Spectrum Sharing Frameworks

There are two dimensions along which spectrum use can be categorized. The first has to do with whether exclusive use relates to licensed or unlicensed spectrum bands. There is considerable evidence that the non-exclusive use of unlicensed spectrum has huge economic value. Recent estimates place the value created by current applications of unlicensed spectrum at $16–37 billion a year in the US alone (Milgrom et al. 2011). Sharing of licensed spectrum bands is also generating enormous value.

The second dimension along which sharing agreements can be categorized is the geographical span of the shared spectrum. Some sharing agreements, for instance TV white spaces, span a wide geographical area. We call these macro-cellular sharing agreements.

Table 7.1   Frameworks for non-exclusive use

	Macro-cell		Micro-cell	
	Equal access	Tiered access	Equal access	Tiered access
Unlicensed	Community Wi-Fi (both equal and tiered access)		Wi-Fi Hotspots (both equal and tiered access)	
Licensed	Spectrum sharing	Dynamic opportunistic spectrum access TV white spaces	NA	NA

Source: Authors' framework.

Others are limited to specific locations like airports or homes. These are called micro-cellular sharing agreements. In some cases, there may also be a hierarchy of use, with graded levels of access rights (that is, primary, secondary, and tertiary). In others, all users may have equal access. The various models of spectrum sharing are given in Table 7.1. The sections that follow elaborate on the specific sharing agreements involved.

Within some of these frameworks, users may need licences for non-exclusive use, while in others no licences may be required

Wi-Fi

The story of Wi-Fi, a WLAN technology standard that ensures connectivity between devices, is the most important example to date of the enormous benefits unleashed by unlicensed spectrum. Just as cellular networks allow the same frequencies to be re-used in different geographic cells, the Wi-Fi technology allows multiple low-power devices to make intensive use of spectrum using suitable transmission methods.

Wi-Fi has its origins in the FCC's 1985 decision to open up 2.4 GHz and 5.8 GHz frequency bands for unlicensed use. These bands previously had limited use for unlicensed devices such as microwave ovens. Earlier WLANs used proprietary equipment and technologies which implied that equipment from one vendor could not communicate with equipment from other vendors. In the late 1980s, vendors collaborated with the IEEE in an attempt to establish a common standard. The basic specifications of the 802.11 standard were agreed upon in 1997. The Wi-Fi standard made use of spread spectrum technology.

Apple introduced the first Wi-Fi compatible laptop in July 1999. Within a few years nearly all laptops were sold with in-built Wi-Fi capability and Wi-Fi access points appeared across college campuses, coffee shops, airports, and private homes as a means of connecting computers to the Internet and to other computers and devices. Commercial vendors also started selling access to the Internet by offering Wi-Fi hotspots.

Today Wi-Fi certified devices include personal computers, printers, video game consoles, streaming devices, security cameras, medical devices, MP3 players, digital cameras, smartphones and tablets. Worldwide about 200 million households use Wi-Fi networks and there are about 750,000 Wi-Fi hotspots. About 800 million new Wi-Fi devices are sold every year. More than 58 per cent of the Wi-Fi devices are mobile devices, exceeding the number of traditional computers. The range of Wi-Fi applications continues to expand.

Community Wi-Fi

Community Wi-Fi refers to the provision of Internet connectivity to large swathes of an area using unlicensed spectrum in a public private partnership mode. Such a service has been provided in Paris, and is being tried in San Francisco (with Google leading that initiative), Minneapolis and other cities. Both access to the network and the backhaul are usually wireless making connection to the power source the only wired point in the access network. This is why Wi-Fi transceivers are often slung on public lamp posts. The service may be free, or provided using a usage fee or advertising model. Use of such networks has so far been shown to be light.

Wi-Fi Hotspots

One area that has seen considerable innovation is the integration of Wi-Fi networks with cellular networks. Wi-Fi offloading is a method by which the traffic is diverted from the carrier's macro-cellular network to a localized Wi-Fi network operating in the unlicensed band, installed typically in homes, enterprises, or public locations, thus relieving the licensed spectrum. Such Wi-Fi hotspots can be deployed by an owner of the venue as a 'private hotspot' (for example, homes, office premises, cafes and restaurants such as Starbucks and Costa Coffee, and hotels); or by ISPs as 'public hotspots' typically in areas such as airports and malls; or by mobile operators, either by themselves or in partnership with ISPs, as 'carrier Wi-Fi hotspots'. Deployment in the former two cases is fairly common in India (though the number of public Wi-Fi hotspots is low compared to other countries) but carrier Wi-Fi is yet to take off. Wi-Fi offloading is primarily meant for data offloading from the macro network thus relieving scarce licensed spectrum. Wi-Fi hotspots are typically connected to landline DSL-based broadband access networks.

In the case of private or public Wi-Fi, the hand-off from the carrier network often requires human or application intervention. Typically the hand-off requires authentication by the user or the application to the nearby hotspot. Moreover, since the carrier network and the Wi-Fi network do not necessarily collaborate in the hand-off, the IP address needs to be allocated when the device moves to the Wi-Fi zone and the connection needs to be reestablished leading to possible jitters and delays or even disconnects. Moreover, when traffic is offloaded from an operator's cellular traffic to the Wi-Fi network, the same level of security and integrity as applied in the cellular network needs to be applied on the Wi-Fi network as well. Realizing this, standard bodies such as 3GPP are working on carrier SIM-based authentication as one of the options to achieve security and integrity along with seamless mobility between cellular and Wi-Fi networks (Sridhar and Buchi 2012). The SIM credentials are used for authentication in the Wi-Fi network, obviating the need for manual or application intervention. The architecture of the carrier Wi-Fi network elements as per the latest 3GPP Release 10 specifications is illustrated in Figure 7.1.

As can be seen in Figure 7.1, seamless mobility from the macro-cellular network to the local Wi-Fi hotspot is provided based on the International Mobile Subscriber Identity from respective SIM cards.

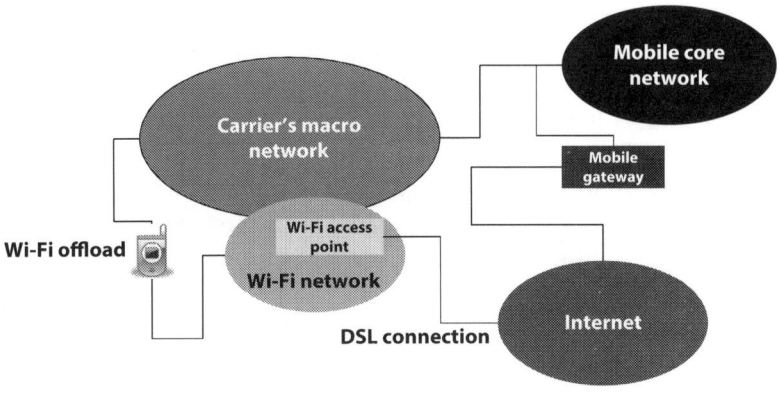

Figure 7.1   Architecture of the carrier Wi-Fi network elements
Source: Authors' framework.

The Wi-Fi hotspot is typically connected through a broadband digital subscriber link (DSL) connection to the Internet.

Operators have been generally reluctant to deploy any Wi-Fi solutions. In 2009, AT&T, the US mobile service provider who bundled Apple's iPhone 3G handsets along with its access service put restrictions on the use of 'Skype over Wi-Fi' and made available only the restricted 'Skype over 3G', to protect their call revenues (Sridhar and Venkatesh 2009). However, there are indications that VoIP is emerging as an important mode for voice communication and independent over-the-top (OTT) VoIP providers such as Microsoft-Skype dominate the VoIP market today. Carriers need to align themselves and provide VoIP offerings. Carrier Wi-Fi provides them a viable alternative to do so.

Similarly while AT&T banned the use of the sling media player that streams broadcast TV content over the Internet on to the mobile as it consumed precious capacity, it was not quick to adopt options to divert such traffic through the Wi-Fi network. In India, such off-loading can be even more critical as each operator gets much lower quantities of spectrum compared to their US and European counterparts.

Today, Wi-Fi bandwidth and speeds are exploding thanks to advancements in coding and multiplexing. The recent IEEE 802.11ac specifications provide a theoretical capacity of 1 Gbps—much more than the long term evolution (LTE) network speeds. The access points conforming to IEEE 802.11u onwards have carrier Wi-Fi specifications implemented and the devices are already available in the market. With US operators adopting them vigorously, the chip-set and device prices are expected to drop dramatically making it a financially viable alternative compared to Femtocells or any other in-building solutions. Future Wi-Fi innovations include in-home video and Wi-Fi direct which is a new technology that supports the connection of mobile devices such as phones, cameras, and gaming devices to each other, without joining a traditional home or office network.

Mi-Fi—A new extension to the existing Wi-Fi

In countries such as India where landline-based broadband connectivity is very poor, there are constraints in deploying Wi-Fi hotspots due to lack of backhaul connections. An architecture of Wi-Fi hotspots to suit this scenario is 'Mi-Fi' (referring to My Wi-Fi, the brand coined and used by Novatel Wireless in the UK) where wireless long haul is provided typically through the 4G-LTE network connection instead of the DSL landline. Thus, Mi-Fi routers are portable and flexible and can connect multiple Wi-Fi enabled devices such as smartphones, tablets, and laptops. Figure 7.2 illustrates the Mi-Fi technology.

Use cases for Mi-Fi deployment include:

(1) Stationary usage where the users connect to the Mi-Fi router at home, office, hotels or even public places for accessing data networks including the Internet. The 4G LTE backhaul provides the needed bandwidth and the licensed frequency spectrum of the LTE network is used for aggregated traffic hence providing the required efficiencies.

(2) Mobile users in public transport such as buses and trains where the Mi-Fi router provides occupants with local Wi-Fi access, connecting them through the wireless backhaul to the Internet. Of course, the handovers across the cells (that is, eNodeBs) need. to take place while a vehicle moves across locations.

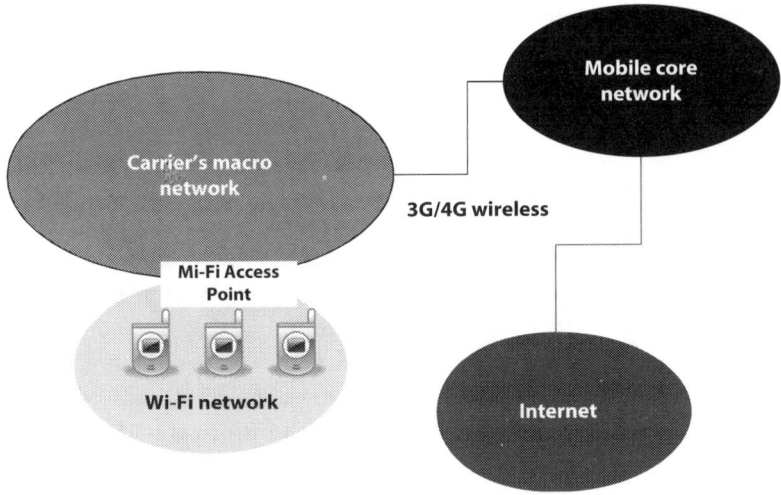

Figure 7.2   Architecture of the Mi-Fi networks
Source: Authors' framework.

(3) Mobile handovers where a nomadic user moves from the macro-cellular network to the Mi-Fi zone where the communication is seamlessly transferred to the local Wi-Fi connectivity provided through the router. This use case is very important in countries such as India where the licensed spectrum allocated to the macro cellular operator is often limited. As soon as the handover takes place the licensed spectrum in the macro network is relieved as in the case of the typical 'Wi-Fi offload' scenario.

There seems to be interest in a large scale deployment of Mi-Fi devices and services, especially from 3G and broadband wireless operators in India.

Spectrum Sharing

Spectrum sharing or pooling is a situation where two or more operators with limited spectrum in a licensed service area pool their spectrum to reap the advantages of trunking efficiencies. They use the combined spectrum in a non-exclusive way using administrative and engineering solutions, including cognitive radio to solve the problem of mutual interference. The Indian regulator has reiterated the need for spectrum sharing, at least in the context of a possible shortage of 2G spectrum (TRAI 2010). It recommends:

(1) Permission for spectrum sharing within the LSA will be permitted initially for a period of 5 years.

(2) Spectrum can be shared only between two spectrum holders. In other words, a non-licensee or licensee who has not been assigned access spectrum as yet cannot be a party to spectrum sharing.

(3) Spectrum sharing will be permitted subject to the condition, inter alia, the total quantum of spectrum (that is, the total spectrum held by the parties in a sharing agreement) as a result of spectrum sharing shall not exceed the limit prescribed in case of mergers of licences.

However, these recommendations have not yet been accepted by the ministry.

While introducing spectrum sharing, one has to consider a number of issues related to the obligations of licensees including spectrum usage charges, roll-out, and emissions limits. With increasing spectrum usage charges common in command and control regimes, the government will need to decide the usage charges to be levied on the two parties in a sharing arrangement. There are three options: continue with the charges prior to sharing; charge both the parties the higher or lower of the two rates fixed prior to sharing; and apply the rate applicable to the combined spectrum block on the assumption that each party uses the whole block. Each of these options comes with its own challenges. Sharing of spectrum among licensees will be facilitated if the annual spectrum usage charges are made uniform for all bands irrespective of the amount of spectrum held.

In the case of sharing of spectrum, each licensee is deemed to enjoy the benefit of the aggregate shared spectrum. For the purpose of assessing the total spectrum holding of the licensee, the total shared spectrum should be counted in the hands of each licensee. In case one of the licensees sharing the spectrum has already fulfilled the roll-out obligations, there should be no further penalties on any of the licensees sharing spectrum. In the case where none of the licensees has fulfilled the roll-out obligations, penalties for unfulfilled roll-out obligations need to be applicable on each licensee separately. In case of sharing, it will be necessary to prescribe responsibility related to frequency, power limits, and interference jointly and severally for compliance of licence conditions of the entire shared spectrum.

In the case of sharing of spectrum bands that are not acquired at market rates, since sharing of spectrum permits a licensee to, indirectly, derive the market value for spectrum acquired at a low price, sharing should only be permitted on payment of 'sharing charges' to the government for the quantity of spectrum shared, in the same manner and of like amount as applicable in case of transfer of the spectrum. Of course, this does not apply to liberalized spectrum obtained using market mechanisms. In the interest of facilitating sharing, when two operators share spectrum, sharing charges should be levied on the smaller of the two spectrum blocks being shared. In case three operators share spectrum, sharing charges should be levied on the smaller two spectrum blocks being shared. Since spectrum sharing arrangements may sometimes stop, the policy may also provide for retention of sharing charges only to the extent levied for the actual period of the sharing on a pro-rata basis, and refund of the difference. In case of a subsequent sale of subsidized spectrum, transfer charges should be payable, pro-rata on the balance period of the spectrum assignment.

Dynamic Opportunistic Spectrum Access

Dynamic spectrum access technologies define a set of protocols for sharing spectrum among different networks. Sharing of unlicensed bands is an example of such a system. Cognitive Radio[1] enables

---

[1] Several definitions for Cognitive Radio can be found in the research domain. The official definition for Cognitive Radio systems in ITU-R developed by ITU-RWP1B in 2009 and published in ITU-R 2009 states that a Cognitive Radio system is 'a radio system employing technology that allows the system to obtain knowledge of its operational and geographical environment, established policies and its internal state; to dynamically and autonomously adjust its operational parameters and protocols according to its obtained knowledge in order to achieve predefined objectives; and to learn from the results obtained.'

secondary networks to use parts of the spectrum that are not being used by the primary network, the one that holds the original spectrum licence.

As opposed to the cooperative spectrum-sharing methods where parties enter into a contract in terms of access rights, in opportunistic spectrum access (also known as non-cooperative dynamic spectrum access), the secondary user does not require permission from the primary rights holder (Chapin and Lehr 2007). The secondary user is thus able to use the best available spectrum band opportunistically, although it does need to make sure that it does not interfere with the primary user. It can work completely on its own trying to sense the environment by itself, collaborate with other secondary users or be assisted by a database running the sensing as a separate service (Weiss et al. 2010).

When looking at opportunistic spectrum access on the level of end users, an analogy can be drawn from the recent emergence of different forms of connectivity clients and mobile network quality databases which enable the user to sense the connectivity opportunities that it has at its disposal (for example, signal strengths of mobile network operator base stations, Wi-Fi access points, or presence of 'white space' spectrum) and to more opportunistically choose the most appropriate network connection. Another recent phenomenon in emerging markets, in particular in India, which bears some resemblance to the opportunistic spectrum access paradigm, is the use of multi-SIM handsets.

Multi-SIM handsets and their relationship to Cognitive Radio

Due to intense competition, mobile operators in India release many tariff plans to attract subscribers. Some plans include only the incoming facility with minimal rentals often referred to as 'life time validity' plans. These plans do not require prepaid customers to recharge periodically as long as they want to just receive calls. The tariff plans differ on the following parameters:

(1) The rental rate
(2) Usage rate
(3) Congestion pricing: different tariffs for peak time and off-peak time

Subscribers optimize their usage based on tariff plans provided by different operators by activating corresponding SIMs. The multi-SIM is also used by subscribers who frequently roam across the 22 licence service areas (LSAs) in India. By having home operator SIMs across roaming LSAs, the subscribers optimize on their roaming charges.

The different operators within the same LSA may not have network coverage uniformly across all areas. This is especially true of new operators who start deploying their networks in high average revenue per user (ARPU) locations first before covering the lower ARPU areas of the LSA. Subscribers who frequently travel across different regions of the LSA, might use a multi-SIM to take advantage of the lower tariff plans of the newer operators while switching to the erstwhile operator network when in an area inadequately covered by the newer operators. The different networks of the operators are also loaded differently. If a subscriber finds an operator's network busy, he or she can switch the SIM to route calls to the alternative operator network.

This is akin to Cognitive Radio where the subscriber uses his or her cognition to switch across networks; or alternatively the intelligent device executes a policy defined by the end-user depending on usage pattern, coverage, and capacity of the networks (Sridhar et al. 2013). It is reported that up to about 40 per cent of all new mobile handset additions in India are multi-SIM ones, though they are relatively expensive compared to single SIM handsets.

TV White Spaces

Another example of opportunistic dynamic spectrum access is in TV white spaces (TVWS). A variation of the ubiquitous Wi-Fi technology that traditionally operates in the 2.4 GHz and 5 GHz, is being tailored to operate in the low frequency TVWS and is popularly known as 'Super Wi-Fi' and is being adopted by the IEEE 802.22af standard. TVWS is defined as unoccupied TV bands in the UHF/VHF bands that can be occupied by secondary (often unlicensed) wireless systems with the condition that the related regulatory and technical requirements are met. The typical use case of communication in TVWS is often associated with wireless coverage of wide areas (for example, suburban and rural areas) where installation of facilities can be costly. In this perspective, the TVWS communication is especially a suitable candidate given its long reaching and high penetrating VHF/UHF signal characteristics (Sum et al. 2011). Taparia et al. (2012) point out that the total capacity associated with it is quite significant. Further, signals in TV bands travel much farther in a cluttered environment than Wi-Fi or 3G signals and have better penetration into buildings with much lower loss. Hence, TVWS becomes an attractive spectrum band to operate. The FCC in the US has allocated TVWS for unlicensed use allowing the operation of cognitive devices on a license-exempt basis. A database system has been set as a primary tool for interference management. On the other hand, Ofcom in the UK has proposed a use of both database system and sensing techniques (Taparia et al. 2012).

It is important to note that this spectrum band is not unlicensed but belongs to broadcasters. Hence, it is privately managed with super Wi-Fi access given as secondary rights. This represents an example of common use of licensed spectrum with tiered access rights at the level of the macro-cell.

The 802.22af standard uses cognitive radio techniques that sense and adapt to the surrounding environment. The IEEE working group is focusing initially on supporting regional area networks in rural areas where other technologies are not commercially viable. These areas have the most available white space channels, the least potentially competing transmissions, and a significant market need for broadband alternatives. Researchers are now proposing technical mechanisms for coordination among independent 802.22af mesh networks, which could further enhance the efficiency of white space utilization. It can be used for long-range rural broadband and for improving short-range coverage.

The recently FCC-approved TVWS database is to be used for secondary uses. Google is one of the TVWS database licensees for tracking and maintaining unoccupied TV channels in select geographic regions in the US. Its open database is available for registered wireless devices to automatically determine the frequencies that they can transmit on after feeding it their encrypted location information. Individuals and organizations can also browse the database to see which bands are available for use in their area.

Case Studies of Unlicensed Use

We now discuss two cases on the application of commons approach to illustrate the challenges and steps involved in the transitions.

Case 1: 3650 MHz for Unlicensed Use in the US

A case study of the allocation of 50 MHz of spectrum between 3650 MHz and 3700 MHz for unlicensed use in the US is instructive (Brito 2006).

This band, prior to being unlicensed, was reserved for satellite communication. Now this band is subject to licensed non-exclusive use. Thus, there can be an unlimited number of licensees.

The FCC has laid down a number of conditions on use including:

- A power limit of 25 watt while higher than the existing unlicensed bands, is lower than what would have been possible in an exclusive licensed zone.
- Only base station–enabled devices are allowed to operate. This precludes broadcasting or mobile to mobile mesh communications.
- Any device for this band must incorporate some type of contention-based protocol; for example, listen before talking.
- The FCC states that all users must cooperate in the selection and use of frequencies in order to minimize harmful interference. They prescribe the use of a common database to register one's base stations before beginning operations.

It is clear from the above case that the management of unlicensed spectrum involves a fair amount of administrative decision-making and that administrative decisions continue to play a major role in the manner in which the unlicensed spectrum gets used. For example, the restriction on power limits in the aforementioned case precludes deployment of normal Base Transceiver Stations manufactured by the network equipment manufacturers for use in mobile networks. The constraints on protocols limit the type of devices and networks to be deployed. This explains the view of the opponents of the specturm commons that the management of unlicensed spectrum requires as much if not more administrative control as a licensed spectrum regime. Any errors in the setting of various parameters such as power limits, type of protocols, networking technology, process requirements and device types can create inefficiencies that defeat the whole purpose on non-exclusive use.

Case 2: Commons of Government Spectrum

As spectrum is moved from command and control or flexible use to a commons, the rights of licensees under the old regime need to be safeguarded. This issue acquires even greater importance in the context of spectrum held by government departments. In this context, the President's Council of Advisors on Science and Technology (2012) in the United States has proposed a model of spectrum sharing with a hierarchy of licence types.

The President's Council's paper argues that since measurements show that less than 20 per cent of the capacity of prime spectrum (below 3.7 GHz) is used even in the most congested urban areas, we need to evolve a new spectrum architecture that enables more efficient use. Today wireless architecture is less commonly being built out for wide-area coverage but is being built for higher aggregate capacity over small areas. Spectrum architecture is moving from a macro-cell to a micro-cell approach. This brings high-frequency spectrum at par with low-frequency spectrum due to the irrelevance of propagation over a large area.

The paper argues that we should move from a narrow-band approach where small swathes of spectrum are licensed to single entities over a large area, to a wide-band approach where wide swathes of spectrum (up to a factor of two in frequency) over localized regions are given to prioritized licensees who share the spectrum in accordance with their place in the pecking order. In order to provide a test case they have recommended that 1000 MHz of Federal spectrum be reallocated for sharing with three categories of licensees:

Federal primary access: Users will register their actual deployments in a database and will be guaranteed protection from harmful interference in their deployed areas. Users will have exclusive use of the spectrum when and where they deploy network assets or in locations where, or times when, underutilized capacity can be put to use without causing harmful interference.

Secondary access: Users will be issued short-term priority operating rights in a specified geographic area and will be assured of interference protection from opportunistic use. However, they will be required to vacate when a user with Federal primary access registers a conflicting deployment in the database. There may be multiple levels of secondary access, either because of payments (for example, auctions) or because of a public interest benefit.

Generalized authorized access (GAA): Users will be allowed opportunistic access to unoccupied spectrum if no Federal primary or secondary users are registered in the database for a given frequency band, specific geographical area, or time period. GAA users will be obliged to vacate once a conflicting Federal primary or secondary access deployment is registered. GAA devices should have the ability to operate on multiple bands and use dynamic frequency selection, so that there is no dependency on access to a particular frequency. Certain bands can also be subject to a device registration requirement.

In the new architecture, entities which manage traffic in small, high-capacity cells could become important players in their own right and provide competition to the wide area network carriers.

Conclusion

As policy-led developments and market innovations demonstrate the feasibility and value of spectrum sharing using a variety of models, important questions arise on the optimal model of spectrum sharing for the future. Table 7.2 summarizes the different frameworks available for both exclusive and non-exclusive use.

Regulatory approaches must allow maximum flexibility of experimentation with new approaches while reserving small bands in high-propagation spectrum bands for sharing in urban areas and much larger bands in rural areas.

Table 7.2 Frameworks for exclusive and non-exclusive use

| | Licensed spectrum | | | | Unlicensed spectrum |
| | Spectrum rights transferred | | Spectrum rights not transferred | | |
	Between operators	Between operators and other entities	Between operators	Between operators and other entities	
Exclusive use	Trading, acquisitions	NA	Intra- and inter-circle roaming (as discussed in Chapter 6)	MVNO, Femtocell operators	NA
Non-exclusive use	NA	NA	Spectrum sharing (equal access)	Dynamic opportunistic access, TVWS (non-exclusive, tiered)	Wi-Fi hotspots, community Wi-Fi

Source: Authors' framework.

References

Brito, J. (2006). 'The spectrum commons in theory and practice', *Stanford Technology Law Review*. Available at: http://stlr.stanford.edu/pdf/brito-commons.pdf. Accessed on 15 February 2013.

Chapin, J., and W. Lehr. (2007). 'The path to market success for dynamic spectrum access technology', *IEEE Communications Magazine*: 96–103.

Milgrom, Paul, Jonathan Levin, and Assaf Eilat. (2011). 'The case for unlicensed spectrum', *Policy Analysis*. Available at: www-siepr.stanford.edu. Accessed on 29 November 2013.

President's Council of Advisors on Science and Technology. (2012). 'Realizing the full potential of government–held spectrum to spur economic growth. Available at: http://www.whitehouse.gov/sites/default/files/microsites/ostp/pcast_spectrum_report_final_july_20_2012.pdf. Accessed on 7 February 2013.

Sridhar, V., and C. Buchi. (2012). 'Is Wi-Fi offloading really good?', *Voice & Data*: 82–83.

Sridhar, V., T. Casey, and H. Hämmäinen. (2013). 'Flexible spectrum management for mobile broadband services: How does it vary across advanced and emerging markets?' *Telecommunications Policy Special Issue on Cognitive Radio* (37): 178–91. Available at: http://dx.doi.org/10.1016/j.telpol.2012.07.008

Sridhar, V., and G. Venkatesh. (2009). 'Let the traffic flow', *Business Line*, 14 September.

Sum, C., Villardi, G., Lan, Z., Sun, C., Alemseged, Y., Tran, H.N., Wang, J., and Harada, H. (2011). 'Enabling technologies for a practical wireless communication system operating in TV White Space.' *ISRN Communications and Networking* 13 (doi:10.5402/2011/147089).

Taparia, A., T. Casey, and H. Hammainen. (2012). 'Towards a market mechanism for heterogeneous secondary spectrum usage: An evolutionary approach', IEEE International Symposium on Dynamic Spectrum Access Networks (DYSPAN), 16–19 October 2012, Bellevue, WA, USA, pp. 142–53.

Telecommunications Regulatory Authority of India (TRAI). (11 May 2010). 'Spectrum management and licensing framework.' Available at: www.trai.gov.in. Accessed on 25 July 2011.

Weiss, M.B., S. Delaere, and W.H. Lehr. (2010). Sensing as a Service: An exploration into practical implementations of DSA, in Proceedings of the IEEE DYSPAN 2010, Singapore, 1–8. Available at: http://dspace.mit.edu/openaccess-disseminate/1721.1/595345 (accessed on 10 February 2012).

8

Net Neutrality

Allowing broadband carriers to control what people see and do online would fundamentally undermine the principles that have made the Internet such a success.... A number of justifications have been created to support carrier control over consumer choices online; none stand up to scrutiny.
— Vinton Cerf, Google Chief Internet Evangelist and co-developer of the Internet Protocol; Turing Award Winner, 2004

Overview

The Internet has provided content providers and application developers with a huge and growing addressable market with very low barriers to entry. Access has been based on the principle of net neutrality, the ideal that the communication passing through an electronic network would depend on the choices made by content and application developers (CAPs) on the one hand and users on the other and not the intermediary provider of the network. It is believed that this freedom led to a thousand flowers blooming, including Wikipedia, the Arab Spring and companies such as Hotmail, Google, Amazon, and Facebook.

Traditionally, the basic feature of the provision of connectivity was the best-effort paradigm. In a best-effort network, all users obtain best effort services, meaning that they obtain an unspecified variable bit rate and delivery time based on the current traffic load. All end-user service requests demanding network capacity are treated equally irrespective of nature or content. However, while best effort is the stated regime on the Internet, traffic management techniques that allow connectivity providers to manage traffic more extensively and to differentiate packet routing based on content, applications, and users are also in use. Traffic management allows for a wide range of operations such as the construction of fast lanes for certain types of data, the provision of guaranteed network capacity for certain types of users, prevention of access to illegal content, and authentication of customers. Advanced traffic management techniques include the ability to charge application providers and users in a differentiated manner, or to block certain kinds of traffic altogether.

The operationalization of the principle of net neutrality takes on a variety of forms. These include a complete proscription of traffic management practices and/or prohibiting any commercial relations between the CAP and the end user connectivity provider (ECP). Some definitions limit themselves to mandating transparency in disclosing network management practices, not blocking lawful CAPs, and not indulging in unreasonable discrimination (FCC 2010).

The changing hues of net neutrality emerge in the context of exponentially increasing amounts of data and increasing varieties of applications, resulting in repeated demands for capacity enhancement of networks.[1] ISPs such as Verizon, Com Cast, and AT&T oppose network neutrality regulations and claim that such regulations will discourage investments in broadband networks. The logic is that they will have no incentive to invest in network capacity unless content providers supporting bandwidth-intensive multimedia applications pay a premium for heavy Internet traffic. Without such a payment, the resulting degradation of quality will prevent applications needing high bandwidth from emerging and succeeding. Verizon, a leading ECP in the US, has argued that the net neutrality regulation is 'the equivalent of a permanent easement on private broadband networks

---

[1] Available at: http://www.tomsguide.com/us/Verizon-First-Amendment-Fifth-Amendment-Net-Neutrality-FCC,news-15760.html. Accessed on 10 August 2013.

for the use of others without just compensation. Access providers have invested billions in broadband infrastructure on the understanding that they can manage access to network facilities and use those facilities to offer the products that their customers want.'

In contrast, proponents of network neutrality regulations (comprising mostly consumer rights groups and large Internet content companies such as Google, Yahoo!, and eBay) note that the Internet has operated according to the non-discriminatory neutrality principle since its earliest days.[2] To support their claim that net neutrality has been the main driver in the growth and innovative applications of the Internet, they rely on the so-called end-to-end design principle under which the control and intelligence functions reside largely with users at the 'edges' of the network, rather than in the core of the network itself. According to them, this creates an environment that does not require users to seek permission from network owners and thus promotes innovations in Internet applications. Tim Berners-Lee, inventor of the World Wide Web and MIT Professor said: 'The neutral communications medium is essential to our society. It is the basis of a fair, competitive market economy. It is the basis of democracy, by which a community should decide what to do. It is the basis of science, by which humankind should decide what is true. Let us protect the neutrality of the net.'

The provision of connectivity is a 'toll good' in the sense that consumption is non-rival up to a certain threshold of users, and the good is excludable, in the sense that users can be prevented from using the good. The debate on net neutrality relates to the right of the intermediary, that is, the ECP, to discriminate between CAPs, either through traffic management or through commercial agreements, in order to better manage scarce network resources. This right is sought to be exercised in the presence of heterogeneous CAPs that present a variety of demands on the network in terms of dimensions like bandwidth and need for real-time communication. The debate becomes vexed due to the market power enjoyed by the intermediary. For instance, in the US the fixed broadband provider is usually

---

[2] Available at: http://www.google.co.in/help/netneutrality_letter.html. Accessed on 13 November 2013.

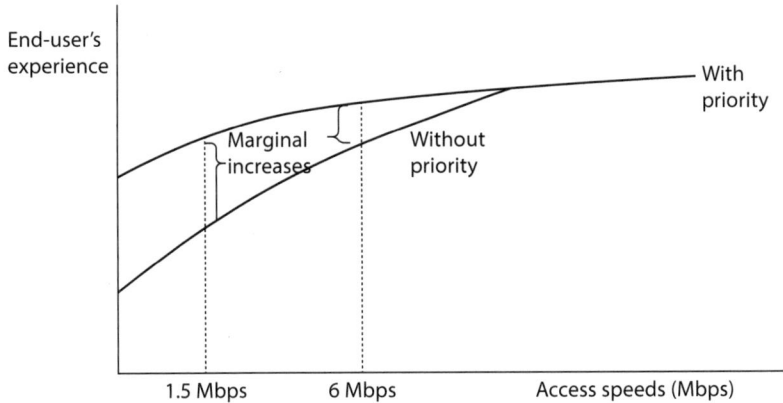

Figure 8.1   End-user experience at different levels of capacity

a monopoly. Hence, the pricing may not merely reflect the scarcity of bandwidth but also the market power of the intermediary.

A further dimension to the net neutrality debate is added by the vertical integration of the ECP and certain CAPs. This gives additional motivation for discriminatory treatment beyond congestion and market power. While vertical integration has been traditionally frowned upon in the context of telecommunications (the breakup of the Bell System being a celebrated instance of dismantling a vertically integrated local and- longdistance service provider), there are some currents of thought that highlight the advantages of vertical integration.

The issue of net neutrality is relevant in a book on spectrum because while there can be different views on net neutrality in the context of a congested network, there can be no doubt that the case for net neutrality becomes strengthened with an increase in the capacity of a network. When the capacity is sufficiently large compared to demand, there is no difference in the end user's experience with and without priority[3] (that is, without and with net neutrality).

Figure 8.1 compares the end user's experience with and without priority.

---

[3] The superiority of the end user's experience with priority at lower capacities is a contested issue in the case of net neutrality. Figure 8.1 ignores the dynamics of innovation in the context of neutral networks.

The capacity of mobile networks, which are growing exponentially compared to fixed line broadband networks (although starting from a lower base), is driven in large part by the scarcity and management of spectrum. The optimal management of spectrum is thus integrally connected with the debate on net neutrality.

To sum up, the issue of net neutrality arises in the context of three inter-related features: network congestion, market power of the ECP, and the vertical integration of the ECP and CAP. Each of these issues is now examined in turn.

Congestion in Networks

The Exploding Demand for Data

The Internet connectivity market has grown from zero to a multi-billion dollar market in 15 years. The traffic conveyed on networks has also been increasing continuously. In 2010, as indicated in Figure 8.2, worldwide IP traffic according to Cisco's estimation stood at 20.2 exabytes per month. Overall IP traffic is expected to quadruple by 2015 to reach 80.5 exabytes per month. As more and more mobile devices are being used to access the Internet, mobile

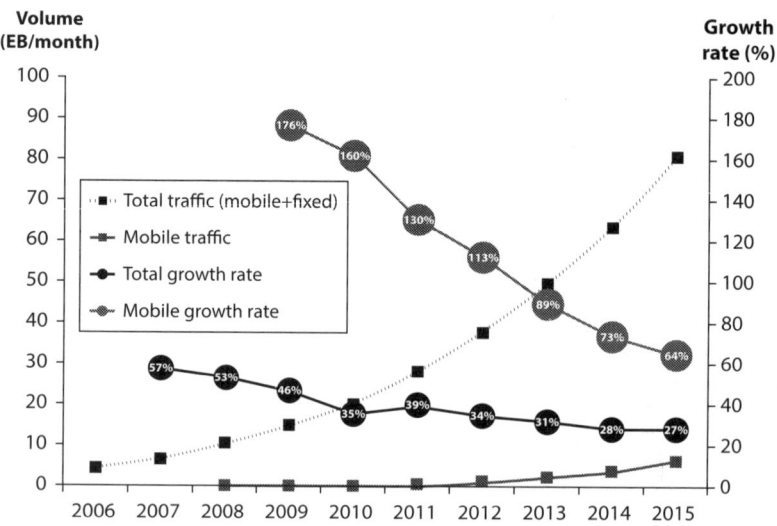

Figure 8.2 Global IP traffic developments (fixed and mobile)
Source: Cisco (Visual Networking Index).

data itself will exponentially grow to about 6.3 exabytes/ month by 2015 as illustrated in Figure 8.3. Further, while 4 billion people on the globe have phone connectivity and 3 billion remain to be connected, there are 50 billion devices that will potentially be wired up in the next one to 15 years. This will place a heavy demand on IP networks.

Simultaneously, Internet applications are becoming more diverse and demanding specific requirements based on their features. The quality of peer-to-peer applications depends mainly on the effective bit rate available whereas delay in packet transmission may be tolerated with only minor effects. On the other hand, the quality of a VoIP call depends on minimization of mouth-to-ear delay. Since different applications require different transmission characteristics, traffic management may be necessary to allow these new applications to appear and grow.

As shown in Figure 8.4, the number of mobile broadband subscribers worldwide overtook the number of fixed line broadband subscribers in 2009 and continues to grow at a more rapid pace.

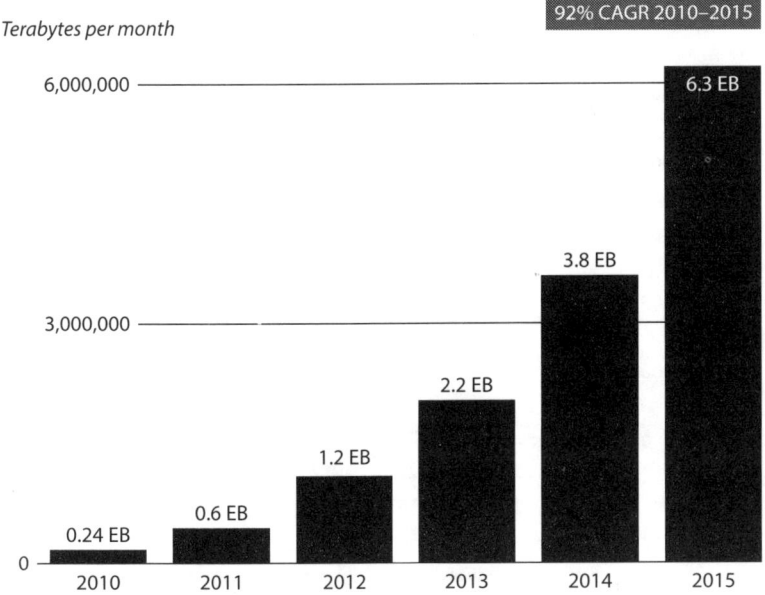

Figure 8.3 Mobile data traffic trends
Source: Cisco (2010).

The 'mobile-only' Internet population will grow 56-fold from 14 million at the end of 2010 to 788 million by the end of 2015 (Sridhar and Hämmäinen 2011). In India, as shown in Figure 8.5, the mobile subscriber base continues to outpace the sluggish growth in fixed line subscriptions. In India the growth of the data is being driven primarily by mobile devices. In emerging economies, including India, wireless access is expected to be the main driver of the uptake of broadband services and the rate of growth of mobile data traffic is expected to continue to be higher than that of fixed data traffic.

Figure 8.5 gives the growth of mobile subscription vis-à-vis landline subscriptions in India.

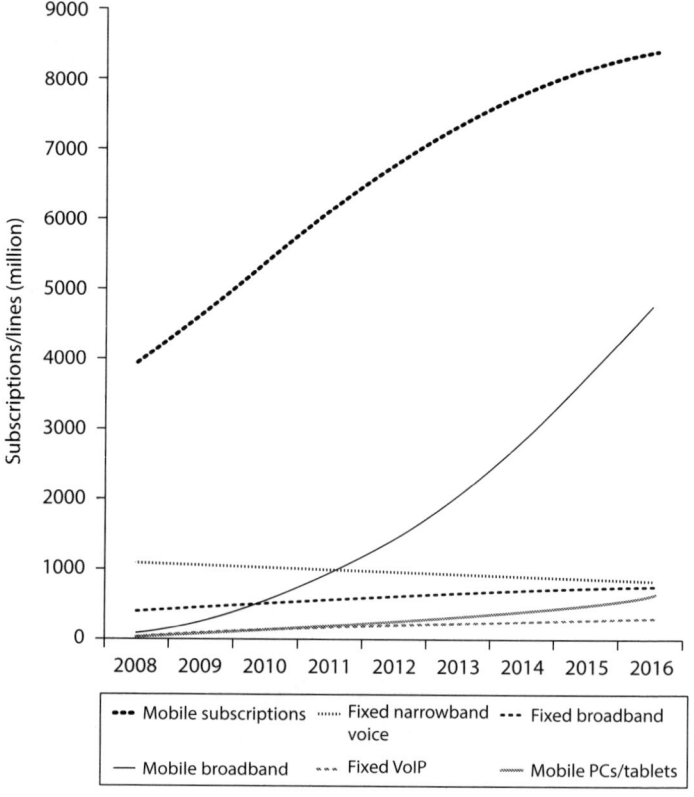

Figure 8.4   Fixed and mobile subscriptions, 2008–16

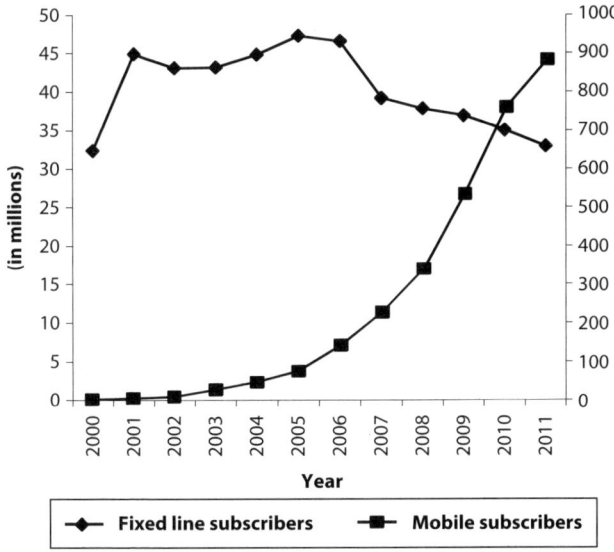

Figure 8.5   Growth of mobile subscriptions vis-à-vis landline subscriptions in India

Expanding Capacity of Networks

The capacity of networks has also been increasing thanks to advances in coding, multiplexing algorithms, and transmission media. Today's optic fibre networks often have capacity in the order of terabits/second. Even wireless networks have been able to attain superior spectral efficiencies and are capable of providing hundreds of megabits/second.

The progress in the transmission capacity of 4G networks has been highlighted in Figures 1.1–1.10, in Chapter 1. The WiMAX IEEE802.16 standard is capable of providing up to 1 Gbps and the FD-LTE standard up to 326 Mbps on the down link and 86 Mbps on the uplink. However, the spectrum available for access networks remains a constraint. This is more so in developing markets, as indicated in previous chapters.

The Economics of Congestion

As a toll good gets congested, traffic management techniques are generally deployed to maintain the quality of service. For example, roads may have a lane reserved for car pooling. Such techniques have also been deployed on the Internet for several decades. The current Internet Protocol version 6 contains header information which allows prioritization of data packets to preserve the quality of service. Greater freedom in traffic management can postpone the onset of congestion. However 'there can be a fine line between reasonable network management, which may be to the benefit of all consumers, and distortion of competition between CAPs (Krämer et al. 2012).

Proponents of freedom in traffic management argue that this freedom prevents the commoditization of access and promotes desirable competition in the network which is good for the consumer. However, supporters of net neutrality argue that increased competition in the network layer can be deleterious if it comes at the cost of a competitive environment among CAPs.

Some of the classic cases involving discriminative practices consisting of traffic management relate to new applications allegedly congesting networks:

- 2007: Com Cast, a cable broadband access provider in the US and Europe restricted the use of certain peer to peer applications on its network.
- 2009: AT&T, the US mobile service provider who bundles iPhone 3G handsets along with its access service decided to put restrictions on the iPhone applications that can run on its 3G network.
- AT&T allowed Sling Player Mobile to stream IP-based video broadcast over Wi-Fi networks not on its 3G network.[4]

In Europe, mobiles services are flat-rate based. Hence, connectivity providers have an incentive to block or degrade applications that consume more bandwidth or consume it in unexpected ways. After all, if the use of the network increases, the connectivity provider's costs increase as well, but due to flat-rate pricing, its revenue stays the same. For a connectivity provider, blocking or degrading selected applications is a quick fix that requires less investment than upgrading the network or devising a non-discriminatory solution. Com Cast's blocking of Bit Torrent and other peer-to-peer file-sharing applications is an example of this type of behaviour. A suggested remedy for distortion of competition on account of traffic management is greater transparency in the disclosure of such practices.

As explained in Chapter 2, a toll good may require a toll to be levied in order to bring about efficient usage. If the toll so determined is not high enough to cover the cost of provision of the toll good, then it may be supplemented by general budgetary outlays if there are positive spillovers involved. Alternatively, the toll may be set at the level of the average cost of providing the toll good for the optimal number of users. In case the toll good in question is connecting two entities, then the toll so computed can be recovered in any combination from the two connected entities. The economics of such 'two-sided markets' is discussed in a later section.

If the users are heterogeneous then they may be charged different prices. High bandwidth users can be charged higher than low bandwidth users. Note that departure from efficiency may even be required in the absence of congestion, if the optimal toll is zero but the cost of provisioning the toll good is positive. However, in the absence of congestion, if a toll is being levied, all users must be charged equally irrespective of their usage characteristics. Similarly all end users must be charged the same price.

The unique issue in the case of spectrum is that without some coordination even a relatively under-utilized spectrum band may result in interference as users in close proximity coincidentally attempt to use the same frequency channel. The usual approach adopted in the case of congestion of mobile networks, in addition to traffic management, has been privatization of spectrum bands, while leaving certain bands free for unlicensed use.

The Market Power of the ECP

In order to understand the issues of market power it is necessary to examine the market structure of the Internet.

[4] AT&T spokesman said about Sling Player Mobile: 'It's absolutely cool [technology], but if we allowed these kinds of services, the highway would quickly become clogged' (Siegel 2009).

Market Structure of the Internet

Most end customers are linked to at most one fixed line and one wireless broadband provider who, in many cases, are integrated with one another. The ECP is therefore referred to as a bottleneck monopoly, in the sense that it is a necessary conduit for any CAP that desires to access any one of its customers. In the case of fixed broadband, there is often a limited or even only a single service provider in a customer's geography. In the case of wireless there may be multiple providers of wireless access. Nevertheless, there is a certain amount of market power enjoyed by the wireless access provider on account of switching costs. These costs go up in countries which do not have mobile number portability or where fixed line networks are poorly developed.

The ECPs have traditionally borne the cost of local access infrastructure deployment to provide broadband connectivity services and passed this on to end users through access and usage charges. In many cases, the ECPs may provide services that address markets traditionally occupied by network operators (and newly contested by CAPs) such as voice services.

The CAPs purchase business hosting and connectivity services from hosting connectivity providers (HCPs). In some cases ECPs and HCPs can be the same entity, however, generally, HCPs enter into 'bill-and-keep' or 'peering' agreements with ECPs to access end users. At present ECPs are mainly charging end users for their broadband Internet access, and not charging CAPs. This situation has been referred to as the zero-price rule. It is to be noted that CAPs do pay for connectivity services to their HCP, and ECPs do charge end users for their bandwidth and do not provide the service free of cost. An illustration of this market structure of the Internet is shown in Figure 8.6.

The Internet has been modelled as a two-sided market where the HCP-ECP combination acts as a platform on which the CAPs and end users interact. An illustration of a two-sided market in which the ISP/telcos is in between is shown in Figure 8.7.

The analytical conclusions of the models of two-sided markets, where the platform provider enjoys market power, are now examined.

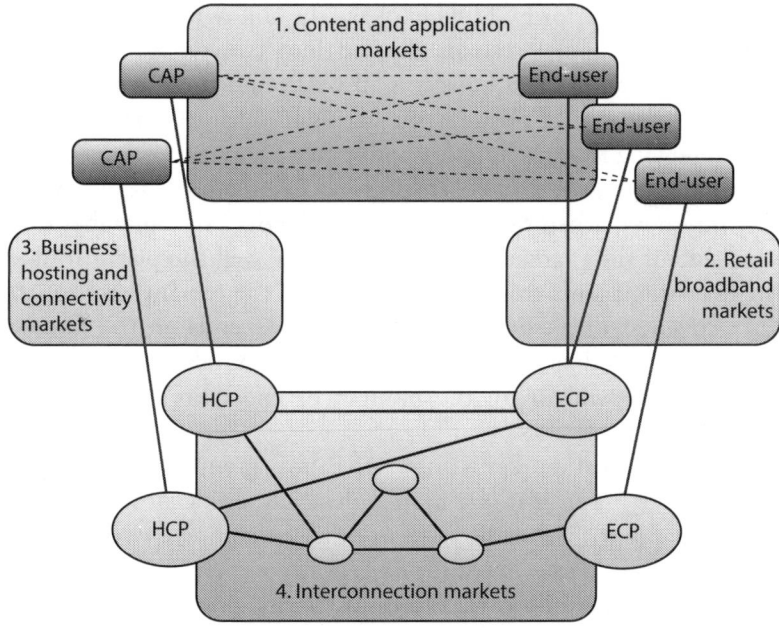

Figure 8.6 Market structure of the Internet
Source: Authors' own.

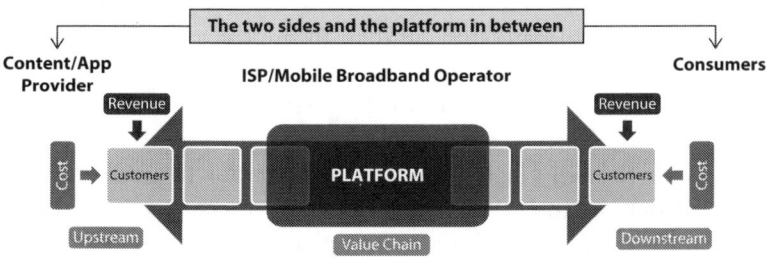

Figure 8.7 Illustration of the two-sided markets
Source: Authors' own.

The Economics of Two-Sided Markets

There are several models of two-sided markets in the specific context of Internet markets that consider the impact of net neutrality regulation on the profits of CAPs and end connectivity providers and on overall social welfare. Most models assume that there is only one or a limited number of ECPs which have the ability to exercise market power. These models can be divided into two types—those with a flat termination fee applied on all CAPs and those with a tiered fee applied to CAPs (besides the access fee charged to end users). The following features hold for both types.

Cross-group externalities: cross-group externalities are present in the Internet market, that is, the benefits enjoyed by a member from one side of the market depends upon how well the platform does in attracting agents from the other side. In the present context, the attractiveness of an ECP to an end user depends on the number of CAPs accessible via the ECP. Similarly, the value of an ECP to a CAP is dependent on the number of end users connected via the ECP.

In a two-sided market with positive cross-group externalities, the intermediary, even if it is a monopolist, has an incentive to reduce platform profit. This is because in order to compete effectively on one side of the market, a platform needs to compete well on the other side. This creates a downward pressure on the prices offered to both sides compared to the case where no externalities exist.

Waterbed effect: a significant feature of the two-sided market of the Internet is that one group of agents, the end users, choose to use only one platform, that is, they 'single-home'. The other group, the CAPs, typically provide content on multiple platforms, that is, they 'multi-home'. In such a market if a CAP wishes to interact with an end user it has no choice but to interact with the user's chosen platform. Thus, platforms have monopoly power over providing access to their single-homing customers for the multi-homing side. This leads to the possibility of high prices being charged to the multi-homing side. By contrast, platforms have to compete for single-homing agents, and their high profits from the multi-homing side are to a large extent passed on to the single-homing side in the form of low prices or even zero prices. This is known as the waterbed effect and has been demonstrated in analytical models like Economides and Tag (2012) under the assumption that CAPs value an additional end user more than the end users valuing an additional CAP. In these models, while CAPs will benefit from net neutrality regula ions that prevent the levy of termination fee, the overall welfare outcome is ambiguous.

Flat Termination Fee

Platform investment decision: Njoroge et al. (2010) model the platform investment decision with flat termination fee and differentiated end users. They find that under net neutrality, the platforms differentiate maximally resulting in a high and low quality platform. Welfare is generally higher with a termination fee because of higher infrastructure investment by the low quality ECP. Better quality leads to higher advertising revenues for the CAP making up for the access fee.

Tragedy of the commons: Musacchio et al. (2009) show that when CAPs have to pay multiple ECPs for access then a situation similar to the tragedy of the commons may arise. ECPs may neglect the negative effect of overcharging on the investment of CAPs and consequently on the revenue of all other ECPs. This negative effect becomes more prominent as the number of ISPs increases and therefore net neutrality, that is, the zero pricing rule, becomes more attractive in this case.

Fragmentation of the Internet: one of the concerns with a termination fee is that, given the large number of CAPs and ECPs, CAPs may be unable to pay the fee to all ECPs and this will lead to a fragmentation of the Internet. The suggested remedy for this is that start-up companies commit to pay a fixed share of revenues for a limited time to the ECPs. This will allow all possible ideas to be introduced to the market with fees only being imposed if the innovation becomes successful.

Tiered Termination Fee

Net neutrality as restricting product differentiation: Hermalin and Katz (2007) examine a situation in which ECPs serve as a platform for connecting content providers with end consumers in a framework of two-sided markets. They consider heterogeneous content providers whose products are vertically differentiated. Without any restrictions, ECPs can potentially offer a continuum of vertically differentiated services, although the ECP is required to provide only one service (a single tier of Internet service) with the net neutrality regulation. They compare the single-service level equilibrium with the multi-service level equilibrium and show that the net neutrality regulation has the following effects: content providers who would otherwise have provided a low-quality service are excluded from the market. That is, content providers at the bottom of the market—the ones that a single-product restriction is typically intended to aid—are almost always harmed by the restriction. Content providers in the 'middle' of the market utilize more efficient and higher-quality service, which favours the net neutrality regulation. Content providers at the top of the market utilize less efficient and lower-quality service than the one that would have been used in the absence of regulation, which obviously favours the discriminatory network. The overall welfare effects of such regulation can be ambiguous, but they argue that the effects are often negative.

Cheng et al. (2011) have developed a game-theoretic model of competition between two content providers in a Hotelling framework. They investigate the effects of net neutrality regulation on ECPs' incentives to expand capacity and also study who gains and who loses as a consequence of the regulation. They assume that the ECP deals with the two competing CAPs non-exclusively. Without net neutrality, it is possible that both CAPs pay the price for priority, but end up being back where they started because no CAP will have an advantage over the other. The ECP stands to gain from the arrangement, as a result of extracting the preferential access fee from CAPs. CAPs are thus left worse off, mirroring the stances of the two sides in the debate. Depending on parameter values, consumer surplus either does not change or is higher in the short run. This is on account of the fact that the quality of the CAPs' offerings are not a variable in the model.

When compared to the baseline case under net neutrality (NN), social welfare in the short run increases if one content provider pays for preferential treatment but remains unchanged if both content providers pay. The incentive to expand infrastructure capacity for the broadband service provider and its optimal capacity choice under NN are higher than those under the non-net-neutrality regime, except in some specific cases. Under NN, the broadband service provider always invests in broadband infrastructure at the socially optimal level but either under- or over-invests in infrastructure capacity in the absence of NN.

Choi and Kim (2010) use a framework akin to Cheng with some important differences. First, they analyse the effects of the regulation on content providers' incentives to provide innovative services. Further, they assume that the ECP deals with CAPs exclusively in that only one CAP can be given priority. In a discriminatory network, both content providers may engage in a Prisoners' Dilemma type of game to receive first priority in the delivery of content and be worse off. The ECP's decision of whether or not it will prefer the discriminatory regime to the neutral network depends on a potential trade-off between its access fee from end users (higher in neutral network as explained later) and the revenue from CAPs through the trade of the first priority (higher in a discriminatory network). Second, the short-run effect of net neutrality regulation on social welfare depends on the relative magnitudes of content providers' cost/quality asymmetry and the degree of content differentiation. In particular, social welfare is higher under the discriminatory regime if the asymmetry across content providers is sufficiently large.

Additionally and more importantly, Choi and Kim study the long-run effects of the net neutrality regulation on the ECPs' investment incentives. They find that there are two channels through which the net neutrality regulation can have impacts on the ECPs' investment incentives: the access fee effect and the rent extraction effect. In the network with net neutrality, capacity expansion speeds up the delivery of content uniformly, thereby enabling the ECP to charge more for access. Similarly, in the discriminatory network, capacity expansion also increases the delivery speed of content and thus allows the ECP to charge a higher network fee. However, because in a discriminatory regime such an effect occurs asymmetrically across different priority classes, one cannot tell unambiguously under which regime the effect of capacity extension is larger.

Capacity expansion also affects the sale price of the priority right under the discriminatory regime. Because the relative merit of the first priority, and thus its value, becomes relatively small for higher capacity levels, the ECP's incentive to invest in capacity under a discriminatory network may be smaller than that under a neutral regime, where such rent extraction effects do not exist. There is an incentive to artificially degrade the quality of the 'best effort' class in order to force more CAPs into the priority lane. This is known as the 'dirty road' fallacy in the net neutrality debate and is invoked by opponents of net neutrality who apprehend a steep fall in Internet quality for the average user.

Overall, the relative incentive to invest under the two regimes is ambiguous. Contrary to ECPs' claims that net neutrality regulations will have a chilling effect on their incentive to invest, the possibility of the opposite cannot be dismissed in this model.

The paper also studies the effects of net neutrality regulation on CAPs' incentives to invest in cost reduction/quality enhancement. Because the monopolistic ISP can expropriate some of the investment benefits made by content providers through the trade of first-priority delivery in the discriminatory network, content providers' investment incentives can be higher under the net neutrality regime. This implies that the ECP's payoff is not necessarily increasing in its ability to extract rents from CAPs when the adverse effects on CAPs' investment incentives are taken into account. As a result, the ECP may wish to limit its ability to extract rent, if such a commitment mechanism is available, to mitigate the countervailing dynamic effect on innovation incentives for CAPs.

Minimum quality of service: overall, in the realm of tiered pricing for CAPs, price discrimination that extracts all or most of the surplus of the CAPs is considered to be undesirable. Minimum quality of service regulation is recommended to avoid the dirty road fallacy. While a non-discriminatory surcharge rule in pricing is considered acceptable, it is not clear how to choose between different forms of non-discriminatory access. For instance, ISPs could charge all services that are in the same class (for example, VoIP) the same fees for data transportation, but services in another class (for example, IPTV) a different price.

Vertical Integration in the ICT Industry

Beyond the market structure of the Internet outlined earlier is the emerging value network (Brandenburger and Nalebuff 2011) of content, applications, devices, access providers of various kinds, and consumers. This presents new issues for net neutrality.

The ICT Value Network

The birth of the Internet and the emergence of next generation IP-enabled networks for mobile broadband have made the vibrant content and application portfolio of the Internet available to mobile telecommunications users. This development transformed the telecommunications service from basic voice connectivity to a 'voice, data, and applications' combination, in which the connectivity provided by the network operator became commoditized. Telecommunications and computing have converged on the mobile broadband device.

Further, iconic mobile devices like the Apple iPhone that provided a compelling user experience along with a bundled music, entertainment, and application package, have become an important fulcrum of the emerging market structure. Apple's development was followed by the setting up of the Open Handset Alliance and the release of the Android open source operating system for mobile devices. Android grew in popularity with handset vendors such as Samsung and the application developer community alike, thanks to its modularity and open interfaces, netting activations of almost one million per day. Finally, powerful CAP brands like Google and Facebook that command the loyalty of hundreds of millions of end users exercise pressure vis-à-vis the other entities in the ecosystem.

The convergence of telecommunications and computing, the rise of the mobile device as a key element of the value chain, and the emergence of transformative Internet applications like Facebook, has undermined the primacy of the network operator. In the wake of new technological possibilities and the growing commoditization of the transmission layer, there has been significant consolidation in the transmission layer with entities combining the roles of ISPs, telcos, cable operators, and direct to home service providers.

Further, entities in the transmission layer have vertical tie-ups with CAPs either directly in the form of cross-holdings with broadcasters, or indirectly through tie-ups with device manufacturers that, in turn, are partnering with CAPs. An example of this would be AT&T's tie-up with the Apple iPhone which comes with an applications store and is integrated with i-tunes.

It is clear that entities other than the connectivity provider impinge upon the interaction between the CAP and the end user, and that the connectivity provider is by no means the only repository of market power. This feature of competition in telecommunications weakens the arguments for regulation focused on connectivity providers alone.

Economics of Regulating Vertical Integration

There is rich literature in economics that looks at the implications of vertical integration on issues of competition. Till the 1970s, anti-trust activity in the US had a long history of looking askance at vertical integration, and encouraging modularity, that is, organizing complements to interoperate through public, non-discriminatory, and well understood interfaces. This was believed to lead to greater innovation in each layer of the market and lower prices, as users mixed and matched components. The outstanding success of the modular approach is the Internet which is based on an open architecture.

However in the last 30 years, the Chicago School has successfully pushed the case for greater tolerance of vertical integration (Farrell and Weiser 2003). One of their arguments rests on the fact that the innovation and quality in the complements market enhances the value of the primary good or platform. One should expect firms to recognize this feature. Therefore, if integration did happen it would only be because the efficiency gains outweighed the gains from free competition in the complements market. This point of view has been referred to as the Internalization of Complementary Externalities (ICE).

Some of the drivers of efficiency gains from vertical advantages that have been pointed out include the internalization of complementary externalities, greater innovation from tight coupling of the complements especially at the early stages of the product life cycle, and the danger of contractual 'hold-ups' in the absence of integration.

However, several concerns about vertical integration persist despite the Chicago School's argument. These include the Baxter's Law or the Bell Doctrine (Joskow and Noll 1999), which states that while free competition in complementary markets is good for a firm, if the firm in question is facing price regulation it may have an incentive to engender and use pricing power in the complementary market. This was the argument made in the case of the participation of the local loop monopolist AT&T in the long distance market in the US in 1982. Other breakdowns of ICE include integration into complementary markets to raise the barrier to entry as any rival will need to enter both the markets in order to compete. A new entrant that focuses entirely on one of the industry silos may be shut out as it is necessary to be part of a value network in order to succeed.

Finally, another caveat to ICE's conclusions is the utilization of a dominant position within one's own value network to drive a hard bargain with network partners. The recent controversy about the working conditions in the Apple supply chain is an example of the issues that can arise (Vascellaro 2012). Another example is Intel's demand for intellectual property licences from its licensees, a practice which attracted the disapprobation of the Federal Trade Commission in the United States (Shapiro 2003).

There are several cases involving discriminative practices related to ECP also being a provider of voice services and discriminating against a CAP that is a competitor in the voice market:

- Madison River Communications, a broadband service provider in North Carolina, US blocked Vonage's Internet telephony service.
- Skype had to restrict its cheap and almost free Internet telephony application designed for the iPhone, to work on the public Wi-Fi network; but not on AT&T's 3G wireless data connection.

A point often overlooked is that the pricing schemes used by connectivity providers for end customers can also amount to discriminatory treatment of content and application providers. An ECP providing a volume discount for high bandwidth users may implicitly be subsidizing the use of high bandwidth applications. On the other hand, a connectivity provider offering a flat rate for unlimited usage but providing lower speed access beyond a certain maximum may be favouring bandwidth light applications.

The Body of European Regulators for Electronic Communications (BEREC 2012) analysed the hypothetical situation in which a connectivity provider blocks the use of VoIP applications by its subscribers over their mobile Internet connection. They examined two scenarios, one in which the connectivity provider has significant market power and the other in which it does not. In their analysis they assume that if a connectivity provider has SMP in the connectivity market it will also have SMP in the market for vanilla voice services as these are usually provided in a bundled fashion. They also treat VoIP and vanilla voice calls as close substitutes.

Scenario 1: VoIP blocking by vertically integrated mobile operator with SMP.

The blocking of VoIP may be done with a view to increasing revenue in the voice market, or to reduce bandwidth requirements. The blocking in this case will have strong effects on end users whose choice will be limited, and on VoIP providers who cannot enter the market. However, when assessing the impact of a SMP operator one must remember that in a multi-node value network, there may be countervailing forces like device manufacturers who command market power in their own right. Indeed device manufacturers like Apple represent two-sided markets as well with CAPs on one side and end users on the other, connected by a device with a high market appeal.

Scenario 2: VoIP blocking by vertically integrated mobile operator without SMP.

If the operator has no SMP, then the availability of non-blocked offers in a competitive market, where transparency and easy switching are effective, reduces the negative impacts of blocking. However, if blocking is implemented by several mobile access providers and becomes widespread, its impact is amplified on both end users and VoIP providers. Note the forces of competition may not lead to a shift to operators who provide unrestricted access as the benefits of unrestricted access including greater innovation may not be fully internalized by industry players. Thus, even in the absence of SMP in the access market a widespread practice of blocking an application can be anti-competitive.

Net Neutrality in Practice

The recently released US Regulations (FCC 2011) state the following with respect to net neutrality:

(1) Transparency: fixed and mobile broadband providers must disclose the network management practices, performance characteristics, and terms and conditions of their broadband services.
(2) No blocking: fixed broadband providers may not block lawful content, applications, services, or non-harmful devices; mobile broadband providers may not block lawful websites, or block applications that compete with their voice or video telephony services.

(3) No unreasonable discrimination: fixed broadband providers may not unreasonably discriminate in transmitting lawful network traffic.

Clearly, the US government expects fixed line broadband service providers with their large data capacity to adhere to the principle of net neutrality in full. Even blocking and prioritizing of traffic are to be eschewed. While clause 2 mandates no blocking for mobile broadband providers of lawful websites or competing content, clause 3 is silent about unreasonable discrimination in the mobile operators' case. Hence the US FCC ruling on net neutrality is partial and does not cover fully mobile operators. Since the mobile broadband service is still evolving the FCC is cautious in mandating full net neutrality for mobile operators thus allowing them to prioritize traffic.

Chile adopted a net neutrality law in 2011. In its final version the law merely states that ECPs cannot arbitrarily block, interfere, discriminate, hinder, or restrict the use of the Internet. Netherlands is the first country to apply net neutrality in full for both landline and mobile operators. However, it is a small country with close to 100 per cent landline broadband penetration. The EU is yet to adopt a firm stand on net neutrality and is working on a case by case basis.

Unfortunately in India, there is no mention of net neutrality in the National Telecom Policy of 2012. Taking into account the high competition in the market, the nascent broadband wireless access provisioning market, and the tiered pricing models already in use, an ex-ante prescription of complete net neutrality may be too premature. It may be appropriate to use a model similar to that if the US where the blocking of competing applications and content is proscribed.

Implications for Spectrum Regulation

The issue of net neutrality becomes pertinent in the context of network congestion. Congestion on mobile networks is driven in large part by the scarcity of spectrum. Therefore, greater efficiency in the management of spectrum can be a key enabler for net neutrality. Indeed, spectrum management is the flip side of the debate on net neutrality.

In the US a study by the President's Council of Advisors states that less than 20 per cent of the capacity of prime spectrum (below 3.7 GHz) is used even in the most congested urban areas.

While spectrum in India may be more clogged on account of much lower average holding of spectrum, there can be no doubt that moving to new models of spectrum management including a greater emphasis on flexible use of licensed spectrum, spectrum sharing, licensed non-exclusive access, and unlicensed access will free up a huge amount of capacity in our data networks.

Conclusions

The analytical models highlight the fact that the impact of net neutrality is subtle and context dependant. Some of the insights are:

- The efficiency of congestion pricing and the admissibility of charging above marginal cost for a toll good, if the marginal cost is less than the average cost.
- The waterbed effect in two-sided markets where one side is multi-homing and the other is single-homing: the intermediary can make high profits on the multi-homing side and pass on some of these profits to the single-homing side.
- Social welfare may decrease with net neutrality if the CAPs are sufficiently diverse.
- Quality may decrease with tiered prici.ng because of the incentive of ECPs to degrade the quality of the best effort Internet in order to force everyone to pay for access.
- The impact of vertical integration depends on the ability of a firm to use a dominant position in one industry to capture market share in a complementary industry versus the need to integrate in order to internalize complementary externalities.

Further, the issue of net neutrality takes on different hues in the context of different relative maturities of fixed and mobile networks in a market. If a country has a dominant means of access, either fixed line as in Bhutan, or mobile as in India, then if net neutrality is established as an important principle it must be applied to the dominant network. In case both means of access are well established in a country, then net neutrality can be applied on the high bandwidth fixed network and need not be mandated on the mobile networks as each consumer can be targeted in an undifferentiated manner by all CAPs using the fixed network. In case both fixed and mobile networks have low penetration, net neutrality may again need to be mandated on both networks, as fixed line networks have capacity, and mobiles are likely to be the chosen means of access.

This chapter attempted to summarize the economic and technology issues involved in net neutrality. However the issue of net neutrality is also fundamentally linked to debates in communications studies about the role of the gatekeeper and the content provider, and to the legal discourse on free speech and private property.

The discourses on net neutrality and media plurality are interlinked because both industries are content-based and therefore, share common principles. As the Internet becomes a channel for the distribution of all kinds of content, including news, one cannot have media plurality unless there is a reasonable environment of innovation and openness on the Internet. Similarly, injunctions on net neutrality will be diluted unless the traditional channels of media also deliver pluralistic viewpoints.

Given the absence of any discussion related to the content provider or significant elements of the value chain like device manufacturers, the implicit assumption of the net neutrality approach is that market power resides in the intermediary but not in the content provider or the device manufacturer. The assumption in the media plurality discourse on the other hand is that market power lies with the content provider and not the intermediary.

The differing approaches of media plurality and net neutrality can partly be explained by the different histories of the traditional media industry and the Internet. The history of the regulation of traditional media goes back to the early years of the twentieth century with roots in print media and terrestrial TV. The role of the intermediary in the delivery of print news and terrestrial TV was negligible. Perhaps due to this reason, not much attention was paid to monitoring the intermediary, even after the intermediary became a vital cog in the value chain with the coming of cable TV, broadband, and DTH.

On the other hand, at the time of the advent of the Internet, the most entrenched entities in the landscape were the intermediaries which provided the dial-up connection—the telephone companies. Telephone companies had traditionally been under the regulatory scanner and so, almost by default, drew attention in the Internet industry as well. Preventing them from taking advantage of their position as a bottleneck monopoly, that is, the sole conduit for the home, was seen as the primary aim of regulation.

Without adopting a doctrinaire position, the following principles on net neutrality can be put forth:

(1) In case of vertical integration, if there is a direct conflict between an ECP and another CAP (for example, Verizon and Skype) and a discriminative practice is observed then the matter needs to be investigated.
(2) There should be enough competition in the broadband market and transparency in traffic management practices.
(3) Spectrum management should evolve in order to integrate models of exclusive use, spectrum commons, and spectrum trading and sharing.
(4) Regulation of net neutrality should be ex-post, rather than ex-ante, that is, the liability should be imposed only when an anti-competitive or restrictive practice is observed.

References

BEREC. (2012a). 'Report on differentiation practices and related competition issues in the scope of net neutrality.' Available at: http://berec.europa. eu/eng/document_register/subject_matter/berec/reports/?doc=1094. Accessed on 8 August 2013.

BEREC. (2012b). *BEREC Report on differentiation practices and related competition issues in the scope of net neutrality.* Available at: http://berec. europa.eu/eng/document_register/subject_matter/berec/reports/1094-berec-report-on-differentiation-practices-and-related-competition-issues-in-the-scope-of-net-neutrality. Accessed on 10 March 2014.

Brandenburger, Adam M., and Barry J. Nalebuff. (2011). *Co-opetition.* USA: Random House LLC.

Cheng, H.K., S. Bandyopadhyay, and H. Guo. (2011). 'The debate on net neutrality: A policy perspective', *Information Systems Research* 22: 60–82.

Choi, J.P., and B.C. Kim. (2010). 'Net neutrality and investment incentives', *RAND Journal of Economics* 41: 446–71.

Cisco. (2010). *Cisco visual networking index: Global mobile data traffic forecast update, 2010–2015*. Available at: <http://www.cisco.com>. Accessed on 15 October 2013.

Economides, N., and J. Tag. (2012). 'Net neutrality on the Internet: A two sided market analysis', *Information Economics and Policy* 24(2): 91–104.

Farrell, J., and P.J. Weiser. (2003). 'Modularity, vertical integration, and open access policies: Towards a convergence of antitrust and regulation in the Internet age', *Harvard Journal on Law & Technology* 17: 85–134.

FCC. (2010). 'Report and order: In the matter of preserving the open Internet broadband industry practices. FCC10-201.' Available at: http://hraunfoss.fcc.gov/edocs_public/attachmatch/FCC-10-201A1.pdfS. Accessed on 10 October 2013.

———. (2011b). 'Preserving the open Internet.' Available at: http://www.gpo.gov/fdsys/pkg/FR-2011-09-23/html/2011-24259.htm. Accessed on 24 November 2013.

Hermalin, B., and M. Katz. (2007). 'The economics of product-line restrictions with an application to the network neutrality debate', *Information Economics and Policy* 19: 215–48.

Joskow, Paul L., and Roger G. Noll. (1999). 'The Bell doctrine: Applications in telecommunications, electricity, and other network industries', *Stanford Law Review* 51(5, May): 1249–315.

Krämer, J., L. Wiewiorra, and C. Weinhardt. (2012). 'Net neutrality: A progress report', *Telecommunications Policy* 37(9): 794–813.

Musacchio, J., G. Schwartz, and J. Walrand. (2011) 'Network economics: Neutrality, competition, and service differentiation', in B. Ramamurthy, G. Rouskas, and K. Sivalingam (eds), *Next-Generation Internet Architectures and Protocols*. UK: Cambridge University Press.

Njoroge, P., A. Ozdagler, N. Stier-Moses, G. Weintraub. (2010). 'Investment in two-sided markets and the net-neutrality debate', Decision, Risk, and Operations. Working Papers Series, DRO-2010-05, Columbia Business School, July 2010. Available at: http://www4.gsb.columbia.edu/filemgr?file_id=735208. Accessed on 10 March 2014.

Shapiro, C. (2003). 'Technology cross-licensing practices: FTC versus Intel', in John E. Kwoka and Lawrence White (eds), *The Anti-trust Revolution* (pp. 350–72). USA: Oxford University Press.

Siegel, M. (2009). 'AT&T spokesman on Clark Howard Radio show.' Reported on CNET. Available at: http://www.cnet.com/news/is-at-t-playing-gatekeeper-to-the-wireless-web/. Accessed on 10 March 2014.

Sridhar, V., and H. Hämmäinen. (15 July 2011). 'Mobile Internet: Indian telecom leading the way', Dataquest, 34–5.

Vascellaro, Jessica E. (2012). 'Audit faults Apple supplier in the Wall Street Journal, 30 March 2012. Available online at: http://online.wsj.com/ article/SB10001424052702303404704577311943943416560.html. Accessed on 15 May 2013.

9

Mobile Partnerships and Alliances

As airwaves become scarce, the spectrum crunch is turning a field of 'haves' and 'have-nots' into a sharply divided set of winners and losers.
—David Goldman, CNN Money[1]

Introduction

Each regime of spectrum management calls for its own set of partnerships to tap its unique opportunities and challenges. The preceding chapters focus on the regulatory challenges of transitioning from one regime to another. This chapter highlights the partnerships that accompany each regime.

Command and Control

In the command and control regime, the government rations the allocation of spectrum and monitors its use closely. Operators with bigger tranches of high-quality spectrum have more bandwidth to satisfy customers' growing demands for mobile phone calls, texts, and Internet usage. This translates into fewer dropped calls and faster download speeds. However, since there is no flexibility in spectrum

---

[1] Available at http://money.cnn.com/2012/02/22/technology/wireless_carrier_mergers/. Accessed on 15 January 2014.

management, the only way for the carriers to accumulate spectrum is through acquisitions. Acquisitions may be undertaken within or outside a licensee's operating region.

Inter-region Acquisitions

Acquisitions outside the acquirer's operating regions are primarily aimed at expanding market access. A classic case in recent times is the $10.5 billion acquisition of Zain in Africa by the Indian operator Bharti Airtel to expand its footprint outside India to the similar but growing market in Africa. The case of Norway's Telenor's entry into the Indian market is similar. Japan's Softbank also intends to enter the US market by bidding for Sprint or T-Mobile. Examples of such inter-region acquisitions are given in Table 9.1.

The other reason for inter-region acquisitions is to get spectrum in areas where the incumbents do not have spectrum, in addition to the associated networks and subscriber base. There are numerous examples of this kind in India. Bharti Airtel expanded its presence and associated spectrum holding across India through inter-circle acquisitions such as JT Mobile in Andhra Pradesh, Karnataka, and Punjab; Skycell in Chennai; Spice Telecom in Kolkata; and Hexacom in Rajasthan. Similarly, the erstwhile Birla AT&T, later christened as Idea, grew by first merging its operations with Tata Cellular; buying RPG Cellcom in Madhya Pradesh and recently acquiring Spice to enter Karnataka and Punjab. Details of such partnerships are given in Sridhar (2012).

Another interesting case is the acquisition of 78 per cent equity by Japan's Softbank, the second largest mobile operator in Japan next to NTT DoCoMo, of the third largest US operator—Sprint Nextel. Softbank acquired the dwindling Vodafone Japan in 2008 and since then has been gaining market share over rivals NTT DoCoMo and KDDI. Of late, Softbank has been bidding aggressively outside Japan, especially in the US for the loss-making Sprint. In an interesting twist, Dish, the largest direct-to-home operator in the US, has also been eyeing Sprint. However, the motivations are different. Softbank sees synergy between its mobile operations in Japan, especially its 4G LTE offerings, and Sprint's US interests. Through its Japanese handset supplier network, Softbank can arrange for bundled handsets, especially LTE enabled ones, to be provided to

Table 9.1 Recent examples of inter-region acquisitions

Year	Partnership	Valuation
In India		
2001	Bharti Airtel acquiring Skycell to enter the Chennai licence area	NA
2006	Telecom Malaysia acquiring 49 per cent stake in Spice Communications	$178.8 million
2007	Vodafone buying 67 per cent stake in Hutchison Essar	$10.9 billion
2007	Russian Sistema acquiring a 10 per cent stake in Shyam Telelink	$11.4 million
2008	Idea Cellular acquiring Spice Communications to enter the Karnataka and Punjab service areas	$6.2 million
March 2009	Japan's NTT DoCoMo acquiring a 26 per cent stake in Tata Teleservices	$2.7 billion
2012	Norway's Telenor acquiring a 49 per cent stake in Telewings Communications Services	
2013	Bharti Airtel in talks to acquire Warid Teleco Uganda	
Outside India		
2010	Bharti Airtel of India acquiring Warid Telecom in Bangladesh	$300 million
2010	Bharti Airtel acquiring African operator Zain Telecom	$10.7 billion
2013	Japan's Softbank's acquisition of 78 per cent stake in Sprint, US	$21.6 billion

Source: Created by the authors.

Sprint in the US market. However, Dish's interest in Sprint mainly lies in augmenting its spectrum holding, as well as providing its video services to Sprint's mobile subscribers. Due to the declining growth in its pay TV DTH market, Dish wants to diversify into the high-growth mobile broadband business. Apart from eyeing Sprint, Dish has been looking at raising its stake in Clearwire, in which Sprint has a 51 per cent stake. Clearwire's spectrum in the 2.5 GHz band is lying underutilized, hence explaining both Sprint and Dish's interest. Dish, in its feverish attempts to increase its spectrum holding, has also bid for the bankrupt Lightsquared for acquiring its L-band spectrum.

Though inter-region acquisitions usually do not attract the scrutiny of regulatory agencies, there can be exceptions. Table 9.2 indicates some of the Indian acquisitions made by foreign telcos which were criticized as the acquisition value was many times greater than the fee paid by the original licensee. This amounted to a windfall gain which could alternatively have been used to fill government coffers (see details in Chapter 10). Most of the licences were quashed due to concerns about the integrity of the process used.

An interesting case is that of Aircel which went on a $3.2 billion pan-India expansion plan after Maxis Communications of Malaysia took over. However, due to poor execution, it found itself in a cash crunch by 2010. It then sold off its tower business to GTL and used the revenues from that transaction to bid aggressively in the

Table 9.2   Controversial inter-region acquisitions in India

Year	Partnership	Valuation
2005	Maxis Communications (Malaysia) acquiring stake in Aircel	$800 million
2008	Etisalat (UAE) acquiring 45 per cent stake in Swan Telecom	$900 million
2008	Telenor (Norway) acquiring 60 per cent stake in Unitech Wireless	$1.3 billion
2008	Bahrain Telecom (Bahrain) acquiring stake in S Tel	$1.25 billion

Source: Authors' own.

3G and BWA auction in 2010. However, the method of allocation of licences to Aircel during 2003–08 on a fixed-fee basis came under scrutiny and cases were filed against the promoter of Maxis in 2011. By 2013, due to weak leadership and lack of a coherent strategy for its BWA roll-out, Aircel found itself carrying a debt burden of about Rs. 24,000 crore.

Intra-region Acquisitions

Through intra-region acquisitions, the acquirer who already has spectrum intends to accumulate more spectrum, acquire a customer base, and reap economies of scale and scope. According to Stanley Sigman, the CEO of Cingular in the US, one of the main reasons for him taking over AT&T Wireless in 2004 in a whopping $41 billion deal was to combine the assets of these two companies to take advantage of economies of scale and scope to be the 'best in class' (Sridhar 2006). The recently barred acquisition of T-Mobile by AT&T Wireless for $39 billion would have allowed AT&T to consolidate its spectrum asset to get captive subscribers of T-Mobile, and to bring in economies of scale to push ahead of Sprint.

Another interesting case is that of Lightsquared, a venture-backed company in the US, that controls a block of the United States spectrum (1525 MHz–1559 MHz) in the L-Band. It received FCC authorization in November 2004 to use this spectrum to build a nationwide 4G-LTE wireless broadband network integrated with satellite coverage. It made arrangements with Inmarsat, another major spectrum holder in the L-band, to increase the amount of contiguous spectrum over North America available to both companies. However, the project ran into problems as the satellite signals in this band interfered with GPS signals causing disruption in location-based services. As mentioned earlier, Dish Network, the prominent satellite television network provider is vying to acquire Lightsquared for its L-band that will complement Dish's Ku band for satellite-terrestrial network coverage.

However, unlike in inter-region acquisitions, intra-region acquisitions are carefully scrutinized for significant market power. An interesting case is that of AT&T and T-Mobile in the US. The proposed merger was considered as limiting competition and hence disallowed by the Department of Justice. The main motivation for Deutsche

Telekom, T-Mobile's parent company, was the acquisition of enough spectrum to roll out a countrywide 4G-LTE network. After the failure of the merger, AT&T had to fork out over 7 MHz of spectrum to T-Mobile as compensation for the failed merger and T-Mobile was suddenly in business. In another case of interest, FCC and the Department of Justice approved the purchase of 20 MHz of Advanced Wireless Spectrum (AWS) by Verizon Wireless from SpectrumCo, owned by a consortium of cable companies such as Comcast and Time Warner. The purchase effectively doubled Verizon's spectrum holding in the AWS-1 spectrum band (1710 MHz–1755 MHz and 2110 MHz–2155 MHz). In the UK, Orange and T-Mobile merged to form Everything Everywhere to provide seamless coverage using their shared infrastructure.

The acquisition of Spice in Karnataka by Idea Cellular is similar. The case dates back to 2008 when Idea Cellular acquired Spice Telecom's operations, including licences held by Spice for six circles, two of which (Punjab and Karnataka) were commercially operational. Idea too had been awarded licences for all of those six circles and had spectrum in five, including Punjab and Karnataka. This overlap of licence areas was the root cause of the problem. The Delhi High Court disallowed the merger in 2011 for the following reasons:

(1) Telecom licence norms did not allow any mobile service operator to own a stake of more than 10 per cent in another operator within the same service area.
(2) Merger of companies which do not own overlapping licences can be allowed if there are at least three operators in each single service area.
(3) Permission for merger can only be granted to a company after three years of the grant of a licence, which is the lock-in period.

While the first two reasons are related to market power, the third is related to the roll-out obligation. However, after the quashing of Spice's licences (as indicated earlier in this chapter and elaborated in Chapter 10), these conditions did not hold good and the merger was approved. However DoT slapped a penalty of Rs. 600 crores for the violation of the first rule during 2010–12.

The important point is that the rules on M&As, especially as related to combined spectrum holding and market share, are much stricter in the case of intra-region mergers. The competition regulator of the telecommunications sector often sets an upper bound on both the market share of the combined entity as well as the percentage of spectrum held. For instance, the M&A regime in India currently allows the merged entity to have up to 50 per cent of market share but only 25 per cent of the total spectrum. It is necessary to set two bounds because an entity with a low market share but high spectrum share can exercise significant market power in the future.

Sometimes, as is the case in India, regulators set a high cap on market share but a low cap on spectrum share. The criteria chosen by the regulator are perhaps intended to reward companies which utilize spectrum efficiently and thereby have a large share of sub-scribers with a low share of spectrum. However, such regulations can prevent desirable mergers from taking place. In a densely populated market it is not feasible to have a high percentage of subscribers with a low percentage of the allotted spectrum. In less dense circles this may be possible. But such circles are 'coverage-constrained' and not 'capacity-constrained', that is, the bottleneck is physical infrastructure not spectrum. Therefore, mergers for the purpose of acquiring spectrum are unlikely.[2]

Many viable mergers bring together a company that utilizes spectrum efficiently and one that does not utilize spectrum efficiently but possesses a sizable amount of it. Mergers of such companies will enable spectrum to be utilized more efficiently. Under recommendations of the kind proposed in India, feasible transactions will be unlikely to occur as they bring together super-efficient operators who would rather compete with each other. The transactions likely to occur, on the other hand, will often be infeasible. Hence, upper bounds on spectrum and market share should be similar in magnitude.

In case an M&A transaction takes place during the lock-in period, similar considerations as in the case of trading and sharing of spectrum would apply. Hence, M&A should only be permitted on payment

---

[2] In such circles, there is a case for allocating some of the licensed spectrum for the commons after which there will be parity between share of licensed spectrum and market share (see Chapter 2 for details).

of 'merger charges' to the government for the quantity of spectrum shared, in the same manner and of like amount as applicable in case of transfer of the spectrum.

Partnerships for Sharing Passive Infrastructure

In the command and control regime, network operators might enter partnerships to share their passive infrastructure, especially in highly fragmented markets. As discussed in earlier chapters, in a fragmented market, each operator gets a small share of spectrum necessitating spectrum re-use. Spectrum re-use requires installation of passive infra-structure including cell towers and the associated electronics. Setting up of towers, especially in urban areas, is expensive. In emerging economies, the towers have to be equipped with back-up diesel gen-erators due to inadequate grid power. In many countries, including India, the mobile industry is one of the major consumers of diesel, a polluting fuel. In developed countries there are often regulations for putting up towers due to perceived health hazards. Hence, it is in the interests of policymakers in both emerging and developed countries, to encourage tower sharing. In India, policies in the initial stages of the mobile industry did not allow infrastructure sharing to encourage green field deployment. However, later passive infrastructure sharing (for example, buildings, towers, unused fibre optic cables) was permit-ted to enable operators to minimize cost and effective utilization of infrastructure. In 2008, active infrastructure sharing, limited to antennas, feeder cables, BTSs, Radio Access Network (RAN), and transmission systems was introduced.

On account of commercial drivers and regulatory openness, a number of partnerships emerged. Most operators in India hived off their tower businesses into separate entities following which partner-ships between these tower units started occurring. The three large private operators, namely, Bharti Airtel, Vodafone, and Idea entered into tower sharing arrangements and created a joint venture tower management company called Indus Towers. The tower arm of Tata Teleservices merged with Quippo Towers to form Viom Networks. GTL infrastructure, a standalone tower company, bought Aircel's tower business to enlarge its presence. The growth of the tower business in India attracted American Tower Corporation to enter the market by

Table 9.3   Major partnerships for tower sharing

Time period	Partnership	Valuation
2008–09	Bharti Airtel, Vodafone, and Idea Cellular formed Indus Towers	NA
January 2009	Quippo Towers acquired Wireless Tata Tele Info Services (tower arm of Tata Teleservices)	$2.8 billion
December 2009	GTL acquired tower business of Aircel Cellular	$1.8 billion
March 2010	American Tower Corp. acquired the tower unit of Essar Telecom	$435 million
June 2013	Reliance Jio signed agreement to use the towers of Reliance Communications	About $2 billion over 10 years of sharing agreement

Source: Authors' own.

acquiring Essar Telecom's tower unit. A recent development that has attracted much attention is the agreement inked by Reliance Jio— the broadband wireless operator belonging to Mukesh Ambani and Reliance Communications owned by Mukesh's brother Anil Ambani for Rs. 12,000 crore. Reliance Jio which won 2300 MHz spectrum in the 2010 auction will use Reliance Communication's existing tower infrastructure to launch its mobile broadband services. This is expected to help Reliance Communications pay-off its debt, reduce the capital and recurring expenditures for Reliance Jio, and cut lead time for service launch. Table 9.3 gives details of partnerships in this area.

Flexible Spectrum Management

In a flexible spectrum regime, as indicated in Chapter 6, secondary markets for the trading and leasing of spectrum will emerge. While trading involves a transfer of spectrum rights to the buyer, leasing, either to another network operator or an MVNO or a Femtocell operator, requires the original licensee to pay the role of a spectrum manager. In that sense, leasing is a kind of partnership between the lessor and the lesee.

Inter-region Leasing

In case of inter-region leasing, the operator does not have licence and/or the associated spectrum to provide access services in a particular region and hence has a roaming contract with the operator(s) in that region to provide services to its subscribers. There is often an inter-operator wholesale reciprocal roaming agreement determined between operators on the basis of which retail roaming charges paid by the subscribers are fixed. In general, there are two types of retail roaming charges levied by operators: (a) roaming rental and (b) usage-based roaming surcharge. The roaming agreement also enables exchange of information between these networks for authentication and billing purposes. While roaming, a subscriber from a particular network (referred to as the home network) uses the mobile network of a different network provider (called the roaming network) seamlessly for communication.

Domestic roaming contributes about 6–8 per cent of the operators' revenue in India. There is every incentive for operators to extract extra rents from subscribers if roaming charges are not regulated. In the case of voice roaming in India, the Telecom Regulatory Authority of India (TRAI) regulation implemented in 2007 ensured no roaming rental be levied by the operators. TRAI also prescribed the maximum permissible per minute charges for roaming calls, irrespective of the terminating networks, and irrespective of tariff plans. These measures and the competition level in each circle have ensured that retail roaming charges are closely aligned to the cost of roaming. In NTP 2012, it is indicated that 'free roaming' will be introduced in which case the roaming calls will be treated much like the normal calls devoid of any surcharges.

In 2013, TRAI reduced ceilings for national roaming calls and SMSes and instituted a new regime for providing flexibility to telecom service providers to customize tariffs for national roamers through special tariff vouchers. It stated, 'the costs associated with national roaming have declined, but not vanished. Hence mandating a fully free roaming regime is simply not practicable at this juncture.'

However, innovations such as multi-SIM mobiles have reduced the relevance of roaming, especially in India. The frequently roaming user typically has two SIMs, one from the home operator and another from an operator in the roaming circle to reduce roaming charges to the bare minimum. It is to be noted that countries such as the US, Australia, and Canada, have had free roaming within their countries for quite some time. In fact, the European Union has drawn up a sliding rate plan to reduce voice roaming charges so as to reduce the difference between roaming and national tariffs to zero by 2015.

The 3G Roaming Pact in India

There can be interesting twists to roaming arrangements between the operators. An instance is the form of roaming deployed by 3G spectrum holders in India through the so-called '3G roaming pact'. The operator who does not have a licence or spectrum in an LSA may sell its own subscriptions and services to customers and provide service using 'roaming' agreements.

The controversial 3G roaming pact is to be distinguished from standard roaming agreements that allow subscribers of an operator holding licence and spectrum in one region to retain uninterrupted service even when not present in their home network. In India, as indicated in Table 6.2 (in Chapter 6), there is disparity in spectrum allocation across different LSAs for wireless broadband services, especially for 3G. Except for a couple of operators, no operator holds 3G spectrum (in 2100 MHz) across the country. Operators who do not have spectrum in certain circles made pacts with operators who had spectrum in those circles to sign up new subscribers. Table 9.4 illustrates how three firms—Vodafone, Bharti Airtel, and Idea Cellular—are able to provide pan-India services (except in Odisha), though none have pan-India spectrum.

The Telecom Enforcement, Resource and Monitoring (TERM) cell of the DoT issued notices to all operators who were involved in the 3G roaming pact, indicating that their operations were tantamount to the deployment of the Mobile Virtual Network Operator

Table 9.4 The 3G roaming pact

LSA	Vodafone	Bharti Airtel	Idea
Delhi	██	░░	
Mumbai	██	░░	
Kolkata	██		
Maharashtra	██		░░
Gujarat	██		░░
Andhra Pradesh		░░	░░
Karnataka		░░	
Tamil Nadu	██	░░	
Kerala			░░
Punjab			░░
Haryana	██		░░
Uttar Pradesh (W)		░░	░░
Uttar Pradesh (E)	██		░░
Rajasthan		░░	
Madhya Pradesh			░░
West Bengal	██	░░	
Himachal Pradesh		░░	░░
Bihar		░░	
Odisha			
Assam		░░	
North East		░░	
Jammu & Kashmir		░░	░░

Source: Authors' own.

model that is not yet allowed. Further, it was argued that the pact undermined the goals of the auction of spectrum. By colluding, a cartel could be formed to pick up spectrum in non-overlapping circles and provide seamless coverage across the circles using the earlier mentioned 'roaming' pact.

The operators appealed against the notice on the grounds that under their existing Unified Access Service Licence (UASL), any access service using any technology could be provided. All three operators who are under scrutiny have UASL in all the LSAs. Hence, they argue that they can provide 3G services in their circles of operation though they may not have spectrum in these circles. It is expected that the issue may be once again referred to TRAI for possible recommendations. This is an issue that may be faced by any regulatory agency that is trying to move from command and control to flexible use.

Intra-region Leasing

In the case of intra-region leasing, though the operator has a licence and associated spectrum in the LSA, he may not have deployed the access network to enable coverage due to economic reasons. Hence, he enters into an agreement with another operator who has deployed network in that area to allow roaming for its subscribers. This method in general reduces duplication of network elements and allows optimal utilization of scarce spectrum among the operators.

In India, even before 2008 when regulation allowing such partnerships was introduced, operators had started using intra-circle roaming to mitigate the coverage problem, especially in rural and remote areas. Unlike in the case of spectrum sharing, this agreement is static in the sense that the priority list for preferred channels to use while roaming is frozen by the operator acting as a spectrum manager.

For the user, the experience is almost seamless as there are no additional roaming charges levied. From the point of view of spectrum holders, it leads to optimal construction of the radio access network within the licensed service area with agreements on leasing of spectrum across different geographical locations within the service area. Recently Reliance Communications and Aircel agreed to provide intra-circle roaming to provide seamless coverage for their subscribers.

Mobile Virtual Network Operators

In flexible spectrum markets, a non-network operator (also called the Mobile Virtual Network Operator or MVNO) may lease spectrum for a period to provide niche services. Virgin Mobile is an excellent example of a company that has leveraged such partnerships to become a highly successful MVNO in the world. In the US, it runs its operations over Sprint's network and spectrum; in Australia over Optus' network; over the networks and spectrum of T-Mobile and Orange in the UK; and that of Singtel in Singapore, to name a few. Since MVNOs do not own spectrum, they are less regulated. In many countries, only registration is required for starting MVNO operations.

Virgin Mobile entered India in 2011 through a franchisee relationship with Tata DoCoMo, the GSM operator. However, its operations went into dispute as India did not have an MVNO policy at that an time. Though Virgin acted as a reseller and franchisee, it came under government scrutiny and had to abandon its operations.

However, despite their proliferation, exceptions apart, MVNOs have not been able to build robust business models compared to their network operator counterparts. The US is full of MVNO failures, for instance ESPN and Helio. Even the well-known MVNO brand of Virgin Mobile failed in countries such as Qatar and Singapore. Table 9.5 indicates the list of successful MVNOs chosen in the recent MVNO World Congress.

MVNOs include a wide variety of business models including Amazon Kindle that provides e-book downloads in Japan using NTT DoCoMo's network. MVNOs are very active in Europe. There are about 120 MVNOs in Germany that have partnerships with the four network operators—O_2, T-Mobile, Vodafone, and E-Plus. Roughly one-third of the mobile subscribers in Germany are customers of MVNOs. Table 9.6 gives a list of some of the MVNO prominent countries.

Femtocell Partnerships

Leasing of spectrum can also take place between a network operator and Femtocell operators who provide in-building solutions in the licensed spectrum band. As discussed in Chapter 6, Femtocells operate in the licensed band and are deployed as in-building solutions or for street coverage thus relieving spectrum requirements in the macro-cell and providing subscribers with good signal strength for voice and data calls. To enable Femto access, both the wireless spectrum resource in the licensed band and fixed-line bandwidth for wired backhaul are necessary. The integrated operators who have licences for both fixed line as well mobile can easily

Table 9.5 Examples of significant MVNOs in the world

MVNO Name/country of operation	Partnerships with mobile network operators
Poste Mobile/Italy; launched in 2007 by Italy's Post Office—one of the 25+ MVNOs in the country	Vodafone Italy
Virgin Mobile/UK based; has presence in Australia, Canada, France, Poland, South Africa, UK, US, Chile, and Colombia	Optus in Australia; Bell Mobility in Canada; PLAY Mobiel in Poland; Movistar in Chile and Colombia; Sprint NextTel in the US; T-Mobile in UK
The People's Operator/UK; launched in 2012 mainly to cater to price-conscious voice users; 25 per cent of the profit is ploughed back into its parent's charitable venture	Everything Everywhere—the merged union of Orange and T-Mobile
Amaysim/Australia; provider of low-cost 3G services	Optus
Lycamobile/operates in Australia, Austria, Belgium, Canada, Denmark, France, Ireland, Italy, Netherlands, Norway, Portugal, Spain, Sweden, Switzerland, UK, and US	O_2 and Orange in UK; T-Mobile and Vodafone in the Netherlands; Movistar in Spain; Telstra in Australia; and T-Mobile in the US
Smartpinoy/UK; owned by Philippine Long Distance Telephone Co	Vodafone

Source: Authors' own.

Table 9.6   Countries that have a large number of MVNOs

Country	Number of MVNOs
Australia	43
France	65
Germany	120
UK	90
US	149

Source: Prepaid MVNO (2013).

launch Femtocell services in partnership with Femtocell product companies. Operators who have mobile licences only but have spectrum for local access as well as backhaul (for example, microwave or low frequency bands) are also well positioned to provide Femtocell service in the same way. However, the operators who have only mobile licences and spectrum for local access may have to forge partnerships with fixed line operators or cable operators for backhaul capacity (Lin et al. 2013).

Apart from the operators, the network equipment manufacturers also are expanding their product portfolio to include Femtocells. Cisco, a manufacturer of Internet protocol–based access and core network equipment, recently acquired Ubiqusys, a UK-based firm that specializes in small-cell technologies. On occasion, owners of public and private venues, such as hospitals, restaurants, malls and movie theatres, can enter into agreements with operators to lease spectrum and manage Femtocell networks within their venue.

Examples of such partnerships are illustrated in Table 9.7.

In an interesting twist, the Shetland village in Wales, a settlement on the south side of West Mainland, Shetland Islands, Scotland, UK has implemented 'open Femtocell' in the village to provide connectivity to its 300 inhabitants. Vodafone is the operator and this is one of its first installations of open Femtocells. Unlike traditional towers, these Femtocell antennas are fixed on roof tops and provide connectivity through Vodafone's 3G network. This is a community project much like community Wi-Fi networks discussed later in this chapter, and is being deployed in rural, less habituated villages. The operators have interest in these community projects as Femtocells decrease the

Table 9.7   Partnerships in Femtocell deployments

Partners	Nature of Partnership
Mobile operators with Femtocell equipment manufacturers	
Cellcom, regional carrier in the US with Airvana, Airwalk Communications, Mavenir Systems, Starent, Taqua LLC	To provide FemtoCloud platform and other Femtocell services on Cellcom's CDMA networks in specific regions of the carrier's coverage area (2012)
KDDI in Japan with Airvana and Hitachi	To improve coverage and service of the KDDI network (2010)
AT&T with Cisco	To provide Femtocell coverage in its service areas
Other network equipment manufacturers with Femtocell equipment manufacturers	
Cisco acquired Ubiquisys for $310 million in April 2013	To augment its small-cell and Femtocell technology capabilities and product lines
IP access, manufacturer of Femtocell and Pico-cell products, partnered with Qualcomm, the chip designer	To enhance IP access capabilities to develop WCDMA residential and enterprise Femtocell products that use Qualcomm's Femtocell station modem chipset platform
Ikanos Communications and picoChip	Strategic relationship to advance the integration of smart mobile devices into the home network through digital subscriber loop technologies

Source: Created by the authors.

cost of network deployment considerably, while meeting the needs of target users.

Commons

We have emphasized in Chapter 2 that it is the 'managed commons' that are relevant to telecommunications rather than 'open access'.

In subsequent sections, we show that a number of partnerships need to be forged at the 'edges', and within the spectrum commons in order to achieve the goals of efficiency and optimal usage.

Partnerships between Mobile Network Operators and Wi-Fi Operators

Offloading traffic from the macro-cellular network to public or private Wi-Fi that uses unlicensed spectrum is one of the ways of managing the spectrum crunch. Though operators were very reluctant to use this method, of late, thanks to exponential growth in data services, they have had to explore this option. Apart from building their own carrier-grade Wi-Fi networks for offloading and for providing Internet services, the operators have also started partnering with Internet service providers, venue owners (for example, malls and hotels), and governments. An excellent illustration of this type of partnership involves British Telecom as illustrated in Table 9.8.

Orange in France has adopted the model of building its own Wi-Fi networks. The operator has deployed about 40,000 hotspots in the premises of its enterprise broadband customers that include transport hubs, hotels, stadia, and cafes. Partnerships with enterprise customers enable Orange to provide its subscribers Wi-Fi access inside their business premises, thus helping it to manage spectrum efficiently. The same is the case with AT&T Wireless that is building its own Wi-Fi hotspots across the US, especially in dense locations such as Times Square in New York City. The $275 million AT&T paid in 2008 to acquire Wayport—a leading provider of managed Wi-Fi services in the US, is paying off now with the additions of thousands of Wi-Fi hotspots in the country.

On the other hand, some of the Wi-Fi operators apart from selling retail Wi-Fi access, have started partnering with mobile network operators to provide wholesale Wi-Fi access. Boingo is a pure play Wi-Fi operator that owns, manages, and operates a global Wi-Fi-Hotspot footprint, a majority of which comprises of major airports in North America, Europe, and Asia, with the rest in stadiums, malls, and restaurants. Boingo sells both wholesale and retail Wi-Fi access. For its wholesale business, Boingo has partnerships with mobile operators, including Verizon and South Korea's KT and LGU+.

Table 9.8   Partners of British Telecom for Wi-Fi access in countries around the world

BT's Wireless Partner Network	Countries
Beeline	Russia
Boingo	Canada, Puerto Rico, US, UK
Chunghwa	Taiwan
Hub	France
Internet Solutions	South Africa
KeZone	UK
Kubi Wireless	Spain
Maxis	Malaysia
NTT Communications	Japan
Orange France	France, UK
Portugal Telecom	Portugal
SFR	France
Star Hub	Singapore
Swisscom Hospitality	Austria, Belgium, Croatia, France, Germany, Hungary, Italy, Luxembourg, Netherlands, Poland, Portugal, Republic of Ireland, Romania, Spain, Switzerland, Turkey, UK
Swisscom Mobile	Switzerland
Tata Indicom Wi-Fi	India
Telefonica	Spain
Telenet	Belgium, Luxembourg
Telia Sonera Homerun	Denmark, Finland, Paraguay, Sweden, Thailand, US
Tomizone	Australia, India, New Zealand
T-Mobile International	Austria, Czech Republic, Netherlands
T-Mobile International (Germany)	Germany
T-Mobile UK	UK
T-Mobile USA	US
VEX	Argentina, Brazil, Portugal

Source: Available online at: http://www.btwifi.com/find/roaming/all-partner-networks.jsp. Accessed on 22 November 2013.

Townstream, another pure play Wi-Fi operator that built close to 800 open Wi-Fi hotspots in Manhattan in New York City is exploring wholesale Wi-Fi deals with operators such as AT&T and T-Mobile.

Even the MVNOs have hooked on to Wi-Fi and have positioned themselves as superior in providing Wi-Fi offloads. Interestingly, Your Karma, the MVNO that provides Wi-Fi access in cities such as Los Angeles, Chicago, New York, and Washington, provides the unique small device that acts as the Wi-Fi access point. Once the user gets connected to the hotspot, the data is sent through the 4G network of the mobile network operator, Clearwire in this case. Thus, many devices can be connected to the Wi-Fi hotspots and the users can have broadband connection, even on the move. In India these kinds of devices are available from mobile network operators such as Tata DoCoMo and Reliance Communications that provide local Wi-Fi access points, backhauled by the operators' 3G/4G network. Other examples include Freedom PoP and NetZero that provide VoIP and Wi-Fi access using Clearwire's 4G backhaul network.

Partnerships between Venue Owners and Wi-Fi Operators

The provision of Wi-Fi access at a venue may be initiated by a venue owner. An interesting case is that of Starbucks, the global coffee shop chain, that announced on 1 July 2010 that its stores in the US would offer free Wi-Fi, via AT&T. In case customers run out of distractions on the web, Starbucks is giving them even more reason to sit and browse, offering free online articles, music, videos, and local information through a partnership with Yahoo!. This is a case of a venue owner, in this case Starbucks, that provides Wi-Fi access through a mobile network operator—AT&T. There are a number of business-to-business websites such as hotspotsystem.com that offer matchmaking services between venue owners and Wi-Fi operators. Recently, Starbucks has switched over to Google for its Wi-Fi networks operating at 10 times the previous speed of 1.5 Mbps.

Community Wi-Fi

Apart from partnerships among mobile network operators, Wi-Fi access providers and venue owners, cities, municipalities, and local governing agencies have tied up with Wi-Fi access providers to provide wireless access to citizens. The most ambitious and an early mover has been the city of San Francisco in the US. However, the contract between the city of San Francisco and the Google-Earthlink combine signed in 2007, hit road blocks due to the tenure of the contract, the speed of access, and privacy considerations. Despite this failed attempt, Google is still said to be working on providing free Wi-Fi access to the people in San Francisco. Since then many city municipalities have signed up with Wi-Fi providers. Some are listed in Table 9.9.

Unintended externalities are possible as a result of local governments providing Internet service to their constituents. A private service provider could choose to offer limited or no service to a region if that region's largest city opts for providing free Internet service, thus greatly reducing the potential customer base. This could prevent other municipalities in that region from benefiting from the services of the private provider. At the same time, the smaller municipalities will not benefit from the free service provided by the larger city. If usage of the publicly provided network became heavier than existing private options, overload issues could arise, forcing the municipality to invest more heavily, thus spending more revenue on infrastructure to maintain the existing level of service. This issue could be compounded if private providers begin exiting a market as mentioned earlier. For this reason, some of the local government–initiated Wi-Fi projects have not seen the light of the day.

Table 9.9 Most notable free Wi-Fi offered by cities

City	Service Provider
Mountain View, California, US	Google
Parks in New York City, US including popular spots such as Prospect Park, Battery Park, and Central Park	AT&T
Los Angeles Staples Centre Stadium, US	Verizon Wireless
Taipei, Taiwan	Q-Ware communications
Tokyo subway, Japan	NTT DoCoMo and KDDI
Waterfront areas of Auckland, New Zealand	Tomizone

Source: Authors' own.

Another interesting project is the 'Loon' project, incubated at the Google X laboratory. On 17 June 2013, balloons were piloted in the areas of Christchurch and Canterbury in New Zealand in coordination with local governments that provide Internet connectivity to the rural communities in that region. The balloons were spaced in the stratosphere at an altitude of about 20 km. They are inexpensive compared to satellites that currently provide connectivity in rural and remote areas in both developed and developing countries. Each balloon can cover a ground area of about 40 km in diameter and can provide data speeds comparable to terrestrial 3G networks using the ISM band of 2.4 GHz and 5.8 GHz.[3] Google had to work with the local government authorities to provide this community access.

The Indian Railways has been exploring the provisioning of Wi-Fi at railway stations for quite some time. A breakthrough of some sort has happened with the Railways Minister Pawan Kumar Bansal promising free Wi-Fi at stations and on board trains in his budget speech in February 2013. In a pilot being run on the New Delhi–Howrah Rajdhani Express, the external satellite antenna in two cars of the train are radio-linked to Wi-Fi access points in individual compartments. Bangalore city also announced piloting its free Wi-Fi zones in select areas of its central business district in partnership with an Internet service provider.

Partnerships across the Telecom Value Chain

Technology convergence has enabled provisioning of applications and services such as broadband Internet connectivity, video-on-demand, peer-to-peer audio and video sharing, collaborative gaming, and augmented reality using a wide range of devices such as smartphones, tablets, laptops, PCs, and televisions, most of which use scarce spectrum for communication. As a result, the modular world of computing and the vertically integrated world of telecommunications are coming together to create new models of competition and cooperation in the integrated sphere of computing and telecommunications.

[3] Refer to http://www.google.com/loon/how/ for details.

The setting up of the open handset alliance and the release of the Android open source operating system for mobile devices suggested that telecommunications may be following the open architecture model of the world of computing. However, Apple has always adopted an approach of offering a tightly coupled bundle of content, applications, and service providers along with its iconic devices.

The developments have raised the question of whether a verticalized value chain comprising devices, applications, and connectivity the Apple way or a horizontal chain comprising arm's length relationships between different elements of the value chain with low barriers to entry as in the Android approach is the model of the future.

Recent trends including the making of the Surface tablet by Microsoft—the first venture by Microsoft in the hardware space, the purchase of Motorola Mobility by Google last year in a possible bid to make its own mobile phone, related attempts by Facebook—the social networking firm, to venture in to making a Facebook friendly device, the acquisition of Ericssons' stake by Sony to build devices that integrate content, games, and applications in Sony's ecosystem, and the sudden emergence of interest in Silicon Valley in real silicon, indicate that vertical integration of hardware and software is emerging as the new norm.

Competition in the ICT space is likely to take place between 'value networks'—complementary firms that coalesce in formal or informal partnerships to provide an integrated computing, communications, commerce, and entertainment experience to the user. The notion of a value network differs from older models of strategic behaviour such as the Porter approach on account of the much greater emphasis on cooperation between firms. Indeed, the proponents of the value network have coined the term co-opetition to refer to this mix of competition and cooperation. Figure 9.1 illustrates the wide array of entities that form part of the computing-telecommunications value network.

There are strong complementarities between content and network access services especially as more and more wireless broadband networks are getting deployed. A network is of no value if content cannot be provided; content is of no use if there is no network to carry it to the end user. The service provider can be facility based or non-facility

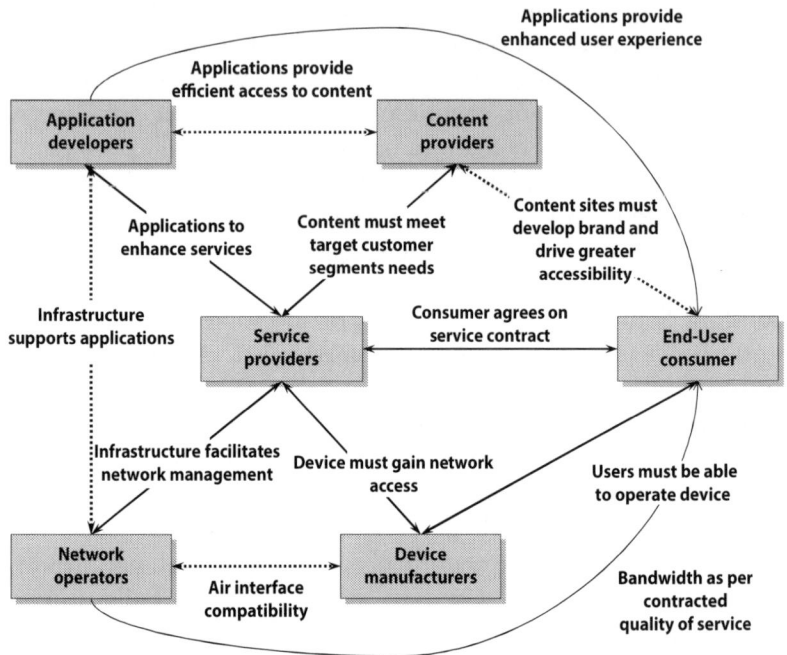

Figure 9.1   The enhanced mobile value chain
Source: Authors' own.

based (for example, MVNOs or ISPs). If they are non-facility based, they depend on network operators to provide the required capacity. The applications, or more broadly the 'platforms', enable delivery of content to the end users to provide 'Quality of Experience' (QoE). Device manufacturers often work with network operators and service providers to enable devices to access content and applications. A classic example is the bundling of iPhones in the US by Apple with select mobile operators. Initially Apple tied up an exclusive agreement to sell its iPhones through AT&T. Google also bundled its first series of Nexus phones with T-Mobile. In India, Reliance tied up with LG and Samsung to bundle the handsets when it launched the CDMA services in the country.

Apart from these types of partnerships, a firm in the value chain might enter into a partnership with another firm to augment its capability to achieve its future technology roadmap. These partnerships are often ways of retaining control over technologies. Google acquired Motorola Mobility for $12.5 billion in 2012 mainly to get access to more than 6,000 wireless patents of Motorola, a domain in which Google did not have any patents. These wireless patents were required for Google to compete against Apple and Microsoft. In emerging economies such as India, Bharti Airtel, the largest mobile operator, struck an outsourcing contract on its entire network management and operations with Nokia Siemens and Ericsson (Sridhar 2006). One of the motivations was transferring technology obsolescence and infrastructure upgrading risk to equipment manufacturers.

In a rather disruptive way, the over-the-top player Skype that offers good quality VoIP and Internet telephony services, launched a partner programme with Wi-Fi hotspot operators. Leading Wi-Fi hotspot providers around the globe including BT Openzone, Fon, Tomizone, and others have signed on to provide Internet access for consumer and business users on-the-go using Skype. With one click, Skype users can connect to the Internet through a Wi-Fi operator partner in over 500,000 hotspots around the globe including 500 airports, 30,000 hotels, and numerous cafes, trains, planes, offices buildings, and convention centres.

Table 9.10 gives a list of some recent partnerships that are aimed at augmenting the value chain or for technology control in the wireless space.

Conclusion

Firms in the telecommunications world are engaging with each other in innovative ways, combining competition and cooperation to navigate the fast-moving technological, commercial, and regulatory waters of the twenty-first century. As we move from the command and control regime of spectrum allocation to more flexible use and towards the commons there emerge newer opportunities and fresher models of partnership between operators, technology providers, content providers, network equipment manufacturers, and handset operators. There is a blurring of the dividing lines between the traditional operators who have spectrum and Internet companies which

Table 9.10 Examples of partnerships across the value chain

Year	Partnership	Reason
2013	Cisco acquired Ubiquisys	To augment its networking capabilities and product range for in-building solutions such as 3G and LTE Femtocells (small cells)
2013	Facebook acquired Parse for $85 million	Parse provides enhanced mobile platform for developers and takes care of data storage and maintenance at the back end. Facebook will exploit mobile data for personalization of its services
2012	Cisco acquired Meraki for $1.2 billion	Meraki produces Wi-Fi gear focused on large campuses and corporations was started in 2006 and emerged from a MIT research project, Roofnet; helps Cisco exploit the Wi-Fi boom
2011	Acquisition of Nortel's 6,000 patents by a consortium of Apple, Microsoft, RIM for $4.5 billion	To gain control in the patent wars in telephony, Internet, search and social networking, mobile phones and networks (GSM, 3G, and 4G LTE); data networking (optical and electrical); and semiconductors
2011	Google acquired Motorola Mobility for $12.5 billion	To mainly get access to more than 6,000 patents of Motorola in the wireless domain in which Google did not have patents; however, required for Google to compete against Apple and Microsoft

2003/2011	Bharti Airtel in India outsourcing its mobile network deployment and management to Ericsson and Nokia Siemens for $724 million	To leverage on the expertise of the network equipment manufacturers to provide state-of-the-art services by transferring technology obsolescence and infrastructure upgrading risk to vendor
2003/2011	Bharti Airtel, the largest private mobile operator in India, outsourcing IT management to IBM for $750 million	To leverage on the expertise of the vendor in developing IT services to meet the needs of the future
2011	Intel acquired Infineon Technologies AG Wireless Solutions (WLS) business for $1.1 billion	To complement Intel's computing platform solutions with Infineon's product portfolio in cellular domain including 2G and 3G; complements Intel's Wi-Fi and WiMax chipset solutions business; helps Intel make inroads into the mobile handsets, consumer electronics and auto infotainment that use the cellular radio service

Source: Authors' own.

are traditionally non-facility based operators. As devices and networks become intelligent, partnerships are required along the length of the value chain, leading to better quality of experience for end users. Nimble regulation in tune with new technological and commercial changes is the need of the hour. Indeed a critical partnership often overlooked in the world of co-opetition may be that between government and industry, a thought gratifying or frightening depending on your point of view.

References

Lin, P., J. Zhang, Q. Zhang, and M. Hamdi. (2013). 'Enabling the Femtocells: A cooperation framework for mobile and fixed-line operators', *IEEE Transactions on Wireless Communications* 12(1): 158–67.

Prepaid MVNO. (2013). 'List of prepaid MVNOs.' Available at: http://www.prepaidmvno.com/mvno-companies/eu-mvno-companies/germany-mvno-companies. Accessed on 18 June 2013.

Sridhar, V. (2006c). 'Mega telecom partnerships.' *Financial Express*, 12 September.

———. (2012). *Telecom Revolution in India: Technology, Regulation and Policy*. New Delhi: Oxford University Press.

10

Hyper Competition and Excessive Spectrum Fragmentation

Case of India

[W]e consider it imperative to observe that but for the vigilance of some enlightened citizens who held important constitutional and other positions and discharged their duties in larger public interest and Non-Governmental Organizations who have been constantly fighting for clean governance and accountability of the constitutional institutions, unsuspecting citizens and the Nation would never have known how the scarce natural resource spared by the Army has been grabbed by those who enjoy money power and who have been able to manipulate the system. In the result, the writ petitions are allowed in the following terms: (i) the licences granted to the private respondents on or after 10.1.2008 pursuant to two press releases issued on 10.1.2008 and subsequent allocation of spectrum to the licensees are declared illegal and are quashed.

—G.S. Singhvi and Asok Kumar Ganguli,
2 February 2012, Supreme Court of India

Introduction

Before the introduction of mobile services, spectrum intended for commercial use in 800 MHz, 1800 MHz, and 1900 MHz was entirely in the control of the national defence force of India. The utilization of spectrum for commercial purposes began with the release of a limited amount of spectrum in 1995. The management of spectrum in the country can be divided into three stages based on different degrees of liberalization of the spectrum market. In the first stage, from 1995 to 2003, the market was relatively nascent but, in principle, auctions were the preferred method for the allocation of spectrum. In the second phase, from 2003 to 2008, the market matured and grew at a rapid pace, but spectrum was allocated through the administered route. The highly controlled spectrum environment unravelled with allegations of the misuse of discretionary power by the government gave way to the third phase where far-reaching liberalization of spectrum is being attempted.

At the very outset, it is important to understand the method and challenges of spectrum allocation in India. Currently, the National Frequency Allocation Plan (NFAP) is the policy document which outlines the allocation of different parts of the frequency spectrum for various purposes. The plan is entrusted to the Wireless Planning and Co-ordination group (WPC) under the Union Ministry of Communications and Information Technology (MCIT). It is revised every two years. All kinds of radio communication technologies and different ministries use the spectrum for various purposes including space communication, radio astronomy, television broadcasting, radio navigation, and mobile communication.

The allocation of frequencies to different uses determines the pace at which the country will upgrade to next generation networks (NGN), the inclusiveness of telecom penetration, and opportunities for indigenous manufacturing. Each of these is dependent on the availability of spectrum within certain specific bands. The ability of the DoT to fulfil its obligations depends on its ability to coordinate between the relevant ministries and negotiate at the International Telecommunication Union (ITU) which aims to harmonize frequencies to reap economies of scale and align with technological progress. Spectrum allocation in India has been mired in controversy with lack of coordination between the Ministry of Defence and the Ministry of Communications and IT. Figure 10.1 represents how the spectrum is shared across the ministries.

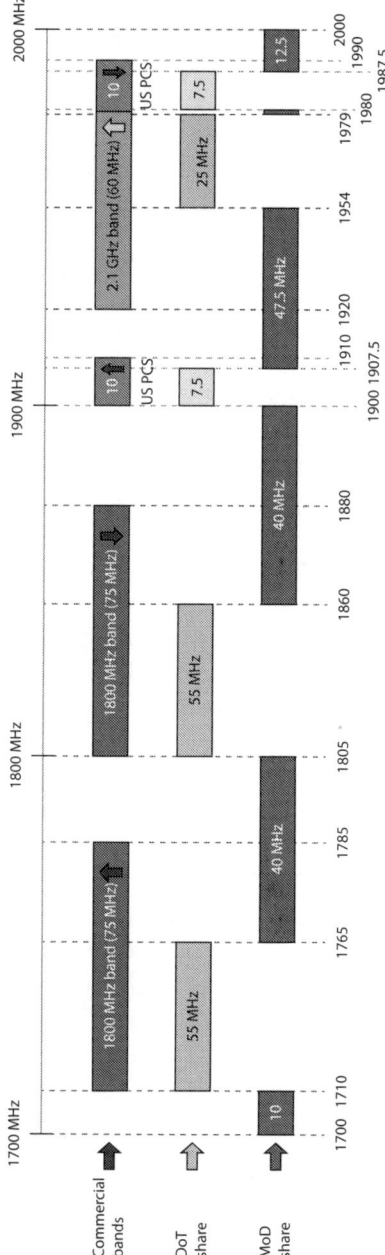

Figure 10.1 Allocation of spectrum between ministries of defence, and communication, and IT in India
Source: Sridhar (March 2013).

Of the 75 MHz spectrum block in the ITU-recommended 1710–1785 MHz/1805–1880 MHz range, only 55 MHz is available for mobile services. Similarly, in the 2100 MHz band, only 25 MHz is available with DoTs while ITU recommends allocating 60 MHz for mobile services. While the harmonized extended GSM band includes 880–915 MHz/925–960 MHz, in India only the GSM primary band of 890–915 MHz/935–960 MHz is used. This leaves only 25 MHz for allocation. Further, 10 MHz has been lost due to possible interference of the lower band with the higher band of 869–889 MHz allocated for CDMA operation. Thus, a major portion of the spectrum used worldwide is not available for mobile services in India. It is in the context of the scarcity of spectrum for commercial services that we analyse the different phases of spectrum management in India.

The First Stage: 1995–2003

To start with, the Indian cellular market adopted a duopoly market structure with licences given to two cellular mobile service providers (CMSPs). GSM was mandated as the technology to be adopted due to its widespread adoption in Europe and the availability of the necessary ecosystem for handsets and network equipment. As in other countries, in India, the CMSPs were licensed to operate in designated geographical operating areas, referred to as 'circles' or licensed service areas (LSAs). In India, there are 23 circles of which there are four metro areas (Chennai, Delhi, Kolkata, and Mumbai). The other circles are normally defined by state boundaries, except in the case of large states such as UP (UP [East] and UP [West]). Some of the smaller states were combined together as a single circle (the North East covering all north eastern states such as Arunachal Pradesh, Manipur, Meghalaya, Mizoram, Nagaland, and Tripura). These circles are categorized as A, B, and C. The categorization is based on the expected revenue potential, with category C circles being on the lower end of the scale. Table 10.1 illustrates the categorization of circles in India for mobile services.

The licensing process for cellular mobile services started in India in 1992. Licences were initially given for the four metros using a 'beauty parade' method. The minimum annual licence fee due

Table 10.1   Circles for mobile services

Metros	Category A	Category B	Category C
Delhi	Maharashtra	Kerala	Himachal Pradesh
Mumbai	Gujarat	Punjab	Bihar
Chennai**	Andhra Pradesh	Haryana	Odisha
Kolkata	Karnataka	Uttar Pradesh (W)	Assam
	Tamil Nadu	Uttar Pradesh (E)	North East
		Rajasthan	Jammu & Kashmir
		Madhya Pradesh	
		West Bengal,	
		Andaman & Nicobar*	

Source: Authors' own.
Note: *West Bengal and Andaman & Nicobar were separate circles initially and were later merged.
** Chennai was merged with Tamil Nadu in 2005.

from the operators over a six-year horizon were specified in advance. The bids were evaluated based on certain pre-qualification criteria including bidders' financial net worth, and prior experience in telecom service provisioning. The financial bids were evaluated based on the monthly rental fee to be charged by the bidders for the first three years. The operator was to pay either the annual licence fee as given in Table 10.2 or Rs. 5,000 per subscriber, whichever was higher. Based on the rentals quoted by the successful bidders, the ceiling rental charge for cellular mobile service was fixed at a low Rs. 156 per month. DoT, the telecom policymaker and licensor in India, planned that about 60 per cent of the licence fee would have to be financed by call charges. Metro licences were issued and the first digital cellular service started in the metros in 1995.

In August 1995, licences were awarded to two CMSPs in each of the other 19 circles using a single-stage auction procedure after a pre-qualification round. The highest bidder was selected for the licence and the second highest bidder was asked to match the winning bid. The licence was given for a period of 20 years. The rentals were fixed at the same rate as metros (for details regarding the licensing procedure, see Desai 2006).

Table 10.2 Licence fee commitments for cellular mobile services

Circles (category)	1st Round (1995)			2nd Round (2001)
	No. of licences issued	Aggregate licence fee bid (in Rs. crore)	Aggregate licence fee paid after revenue sharing in 1999 (in Rs. crore)	licence fee (in Rs. crore)
Delhi (Metro)				170.69
Kolkata (Metro)				78.00
Mumbai (Metro)				203.67
Andhra Pradesh (A)	2	2,002	569.51	102.98
Gujarat (A)	2	3,588	1,020.88	109.00
Karnataka (A)	2	2,786	770.74	206.83
Maharashtra (A)	2	3,315	913.21	188.98
Tamil Nadu (A)	2	1,672	282.91	233.00 (includes Chennai)
Haryana (B)	2	480	116.90	21.45
Kerala (B)	2	1,034	295.06	40.55
Madhya Pradesh (B)	2	10.20	29.12	17.42
Punjab (B)	2	2,532	849.00	151.78

Rajasthan (B)	2	764	217.31	32.24
Uttar Pradesh (W) (B)	2	422	115.90	30.57
Uttar Pradesh (E) (B)	1	812	138.26	45.26
West Bengal, Andaman & Nicobar (B)	1	42	12.24	0.99 (fourth licence issued in 2004)
Assam (C)	1	1.00	0.38	5.02 (fourth licence issued in 2004)
Bihar (C)	2	273	89.49	9.98
Himachal Pradesh (C)	2	30	8.54	1.12
Jammu & Kashmir				1.98
North East (C)	2	4.00	2.42	1.98 (fourth licence issued in 2004)
Odisha (C)	2	178	58.48	5.02 (fourth licence issued in 2004)
Total		**19,945**	**5,490**	**1,658.50**

Source: ICRA (2002), TRAI (2013).

The award of licences through a single-stage auction garnered high licence revenues for the government treasury and reduced the regulatory workload of allocating spectrum. Since the telecom landscape was still nascent, the bidders had little idea how to evaluate market potential. The single-stage auction provided little leeway for the bidders to revise the bid amount based on the available information. Further, there was no requirement for an upfront payment of the winning bid amount. The winners had to pay the bid amount over the licensing period using a deferred payment mechanism. These factors resulted in high bids.

The winner of the auction received a CMSP licence and start-up spectrum of 2 × 4.4 MHz (paired Frequency Division Duplex spectrum assignment) in the 890 × 915 MHz band paired with 935–960 MHz for GSM. Two operators were selected for each LSA. Subsequently the third operator licence along with 2 × 4.4 MHz of start-up spectrum in the same band was awarded to the government operator on a pro bono basis in 2001.

The aggregate licence fee is shown in Table 10.2. Across all LSAs the total bid amount was a staggering Rs. 19,945 crore. However, the euphoria of the auction was short lived. It soon became evident that the bidders had fallen prey to the classic 'winner's curse'. The licence fee payable over the licence period was two to three times the annual fixed service revenue of the incumbent monopoly government operator BSNL (ICRA 2002). The bidding was higher in category A circles and lower in category C circles. The CMSPs were liable to pay their licence fee commitments over the licence period. However, the adoption of mobile phones and their usage did not pick up as expected. One of the main reasons was the low rental (as dictated by the beauty parade based allocation for the metros) and high usage charges (equal to the ceiling price of about Rs. 16 as prescribed by the government). Hence, the licence fee payments were roughly equal to annual revenues of these operators and imposed a heavy financial burden in the initial years of operations.

In 1999, the government realized that most of the operators were not able to pay even the annual fee towards the bid amount. The licence fee payable till 31 July 1999 was as high as 20 times the annual cellular services revenue for the financial year 1999. A study of the financial performance of CMSPs by the regulator indicated that most of the operators accumulated losses till 1999. The accumulated losses were higher for category A circle operators than for operators in category B and C circles, mainly because of higher licence fee payments by category A circle operators. Though the operating revenue in metros was higher for the operators, all operators in metros had also accumulated losses till 1999.

To correct this situation, the government intervened and proposed a migration package for the existing licensees from a fixed licence fee format to a revenue sharing one. The package included an entry fee approximately equivalent to 2.8–2.9 years of the original licence fee commitments (that is, close to 30 per cent of the bid amount) and a continuing revenue sharing arrangement. All the operators paid the arrears and migrated to a revenue sharing scheme, except for one operator who could not pay the entry fee and subsequently lost his licence in UP(E), UP(W), Bihar, and Odisha. Details of the licence fee finally paid by the operators for each circle are also given in Table 10.2. The revenue sharing percentage recommended was 17 per cent of the adjusted gross revenue (AGR) for incumbent migrating CMSPs and included contribution to the Universal Service Obligation fund, research and development, administration, and regulation expenses. AGR was defined as the 'Gross Revenue' accruing to the licensees by way of operations of the cellular mobile service and also included revenue on account of value added services, supplementary services, and sale of handsets plus revenue accruing through resellers and franchisees. Revenue on account of interconnection was naturally excluded as one operator's revenue on this count would be cancelled by another operator's cost.

Allocation of spectrum beyond the start-up spectrum levels was based on availability and justification and attracted additional revenue share as spectrum charges. The contractual rights of spectrum holders were incrementally established through a series of government orders.

The Perfect Auction in 2001

The licensing process for the fourth cellular licence was initiated in 2001. Bids were invited for a 'lump sum' to be paid as entry fee. Subsequently, licences were issued in August 2001, using a three-stage ascending auction procedure. As compared to the single-stage auction used in the first round of licensing, in the multistage simultaneous auction all the licences were put up for bid at the same time and the bidders had an opportunity to bid for as many licences as they wanted and participate in successive rounds of bidding. A simultaneous multistage auction generates more information for the bidders concerning licence values and facilitates the award of the licences to bidders who place the highest value on them (as discussed in Chapter 4). In general, multistage auctions lessen the risks of bidders winning more than they want, and at a higher cost than they would have desired. The winning bid amount for the fourth cellular licences is given in Table 10.2. The reduction in bid amount compared to the first round is evident and is obtained for the following reasons:

(1) The cellular mobile service market had grown and showed a certain maturity level and hence it became possible for the operators to bid realistically considering the market potential.

(2) The three-stage auction procedure allowed the bidders to revise their bids upwards at each stage, thus reducing the probability of a 'winner's curse'.

(3) There was upfront payment required as opposed to the deferred payment as adopted in 1995 that resulted in bidders exercising caution during the process.

Though the government removed the requirement of GSM as the technology to be adopted in the fourth cellular licence guidelines, all the licence winners adopted GSM to make their network interoperable with existing networks. As can be noted in Table 10.2, most of the category C circles were not picked up by any operator in 2001 and were allotted later. As before, the fourth operator licence was issued along with start-up spectrum of 2 × 4.4 MHz in the 1710 × 1785 MHz paired with the 1805–1880 MHz band. In addition to the entry fee, licensees were required to pay a percentage of annual revenue as spectrum charges.

Subscriber Based Criteria for Spectrum Allocation

In 2002, the subscriber-linked spectrum allotment procedure, referred to as Subscriber Based Criteria (SBC) was introduced. This made the release of additional spectrum contingent on the achievement of

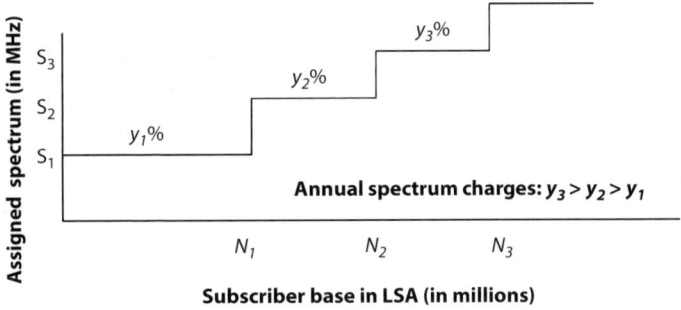

Figure 10.2   Subscriber based criterion for spectrum assignment
Source: Authors' own.

certain subscriber milestones (see Figure 10.2). New entrants get S_1 MHz of spectrum as bundled with licence and pay y_1 per cent of their AGR as annual spectrum charges. Once they cross N_1 subscribers they are eligible to hold S_2 MHz (that is, an additional $(S_2–S_1)$ MHz) and pay y_2 per cent of AGR, where y_2 is greater than y_1. This type of allocation and levying of annual spectrum charges continues until a ceiling of spectrum holding is attained.

The increase in percentage of annual spectrum charges allowed the operators to trade-off the acquisition of extra blocks of spectrum versus the re-use of the current holding of spectrum in congested areas (by increasing the capital and operating expenditure in order to maintain additional towers and BTSs). Since telecom infrastructure was very poor in the country, the initial policy objective was to encourage and mandate private operators who got licence and spectrum to set up the needed infrastructure. Hence, the government, after assigning the spectrum to licensees, expected them to develop infrastructure. Strict roll-out obligations were imposed on the spectrum holders with penalty clauses for non-compliance. The spectrum was assigned based on the principles of maximal usage.

Further, the graph can be shifted to the right or left in accordance with the objectives of the government. In India, the graph was shifted to the right many times, thus tightening the threshold subscriber base for allocation of additional spectrum. TRAI (2007) stated that the spectrum allocation criterion should take into account methodologies such as synthesized frequency hopping and use of advanced

Table 10.3   Spectrum allocation criterion in 2002

Quantum of spectrum allotted	Minimum subscriber base required (in millions)	Annual spectrum charges (as percentage of adjusted gross revenue)
2 × 4.4 MHz	–	2%
2 × 6.2 MHz	–	3%
2 × 8 MHz	0.5	3%
2 × 10 MHz	1.0	4%
2 × 12.5 MHz	1.2	5%

Source: Wireless Planning Committee (2008).

codecs for efficiently utilizing the allotted spectrum. Following this, the government tightened the subscriber base criterion for additional spectrum allocation to existing operators making them ineligible for participating in the new allocation of spectrum.

Tables 10.3 and 10.4 show the original criteria in 2002 and the vastly more stringent criteria in 2008, respectively. For instance, the minimal subscriber base for getting 10 MHz was changed from 1 million to 6.8 million.

This method of spectrum allocation was very different from the methods followed in other countries where a sizable spectrum block (about 2 × 15 MHz) was given to the operators as start-up spectrum (TRAI 2005).

The rationale for adopting a different approach was the scarcity of spectrum due to non-availability from the Department of Defence.

Unified Access Service Licence and Administrative Pricing of Spectrum: 2003–8

This period is characterized by the entry of CDMA operators, the introduction of UASL, and the steep growth in the mobile market on the back of subsidized pricing and high competition.

Unified Access Service Licence

Some basic service operators started using the limited amount of spectrum that they had been given to provide fixed telephony

Table 10.4   Spectrum allocation criterion for additional spectrum allocation in 2008

Quantum of spectrum allotted	Minimum subscriber base required (millions)	Annual spectrum charges (as percentage of adjusted gross revenue)
For GSM services		
2 × 4.4 MHz	–	2
2 × 6.2 MHz	0.5–0.8*	3
2 × 7.2 MHz	1.5–3.0	
2 × 8.2 MHz	1.8–4.1	3
2 × 9.2 MHz	2.1–5.3	
2 × 10.2 MHz	2.6–6.8	4
2 × 11.2 MHz	3.2–6.8	
2 × 12.2 MHz	4.0–9.0	5
2 × 14.2 MHz	5.7–10.7	5
2 × 15 MHz	6.5–11.6	6
For CDMA services		
2 × 3.75 MHz	0.15–0.40	2
2 × 5.0 MHz	0.5–1.2	2

Source: WPC (2008).
Note: *Range across the SLAs.

services to provide mobile services using the CDMA technology. They built up a large subscriber base. Their success indicated that the separation between basic and mobile services was no longer valid. As a result, the Indian government announced UASLs in November 2003 that allowed the operators the use of any access technology to provide any access service (DoT 2013). This paved the way for alternative technologies such as CDMA to be pursued by potential operators. Some of the Basic Telecom Service (BTSs) licensees migrated to UASL and started providing CDMA-based cellular service (for complete details on how UAS came in to existence, refer to Sridhar 2012).

However, the portrayal of CDMA as a spectrally efficient technology compared to GSM by the basic telecom operators backfired when the government used this logic to allot only 2 × 2.5 MHz of start-up spectrum as compared to 2 × 4.4 MHz for GSM. Further, due to limited availability (25 MHz) in the 800 MHz band, the maximum spectrum permissible for CDMA operations was also restricted to 2 × 5 MHz. This constrained the growth of CDMA operations in the country. The minimal subscriber base and annual spectrum charges for different holdings of the 800 MHz are given in Table 10.4.

In 2005, TRAI reviewed the spectrum allocation process taking into account spectrum availability and efficient techniques for utilization of assigned spectrum (TRAI 2005). A maximum of 2 × 10 MHz was being allotted to GSM operators while the world average was about 2 × 20 MHz. Additional spectrum allocation in the 900 MHz band was not possible on account of usage by the ministries of defence and the railways. Hence, any additional spectrum for the first three operators had to be allotted in the 1800 MHz band. However, even the 1800 MHz frequency band was being extensively used by the Air Force and Army networks and could not be released fully to GSM operators. A maximum of 2 × 5 MHz was allotted in the 800 MHz band for the CDMA operators while the world average for CDMA operation was about 2 × 15 MHz. Defence was using the 1900 MHz band and hence this was not available to either GSM or CDMA operators. The 1900 MHz was also being coordinated by the Ministry of Defence for indigenously developed CorDECT fixed wireless local loop technology.

In the light of spectrum scarcity, TRAI (2005) recommended that existing operators be given adequate spectrum before considering allocating spectrum to new service providers especially since 'there is adequate competition in almost all service areas'. TRAI (2006) continued to maintain that there was a shortage of 2G spectrum.

Meanwhile, additional spectrum was also getting vacated by the defence ministry. In parallel with the debate on the shortage of 2G spectrum, there ran a debate on whether the 2.1 GHz spectrum, which could be used for 3G services, was to be treated as an extension of the 2G spectrum, meant to alleviate the shortage, or as spectrum that would be used to provide new services. As long as TRAI was of the view that the 2G spectrum was in short supply, it recommended that the 3G spectrum be treated as an extension of the 2G spectrum and be restricted to incumbents. However, when it came to believe that the 2G spectrum was not in short supply, based on enhanced spectral efficiencies in relation to market demand, then the contrary position was adopted. The authority recommended that 'the Government should not treat the allocation of 3G spectrum in continuation of 2G spectrum'.

TRAI's (2007) recommendation that no cap be placed on the number on telecom access service providers in the country allowed more new firms to enter the market by paying the low fixed entry fee (TRAI 2007). It even allowed CDMA operators to obtain GSM licences and vice versa, that is, become dual technology operators after paying the fixed fee as determined in the 2001 fourth cellular licence. Since licence and start-up spectrum were bundled, the first-come-first-serve basis was used as a methodology for assigning spectrum to the licensees. The new licensees were to be put in the queue for assignment of spectrum as and when available.

Rush to Acquire Mobile Licence and Spectrum Post 2007

Dual Technology Licensing

As of 2006, a company could hold only one UAS licence, and employ one technology, either GSM or CDMA. Reliance Communications (RCL) was the first CDMA operator which decided to join the GSM bandwagon by requesting DoT for spectrum in the 1800 MHz band. This sparked off speculation about the end of CDMA-based services in India. There were two major reasons cited for this move by the Anil Ambani group. The first was the scarcity of spectrum (a total of 2 × 20 MHz available in the 800 MHz band). The second was the discriminatory set of rules laid down by the government for additional spectrum allocation favouring GSM operators. The subscriber criterion was almost twice as stringent for CDMA operators as compared to GSM operators. For example, as per the guidelines, CDMA operators in Delhi and Mumbai were eligible for the sixth carrier amounting to a total spectrum of 7.5 MHz only after they hit 21 lakh subscribers while the GSM operators in these areas were eligible for 15 MHz for the same subscriber numbers. There were indications that more spectrum would be released by the government for GSM operators in the 1800 MHz band. The chances of the government releasing 1900 MHz for CDMA operators seemed remote.

Rules permitting a company to hold more than one UAS licence, and thus use more than one technology were notified in 2007. The factors listed earlier prompted RCL as well as Tata Cellular to opt for GSM as their future growth engine. It was quite an irony that in 2003 the same companies had pressurized the government to permit their entry into the CDMA-based mobile services market under controversial circumstances. The about face of the companies was in part a reflection of the inadequate policy support for CDMA by a government that had gone to great lengths to permit its entry through the UAS licence regime (Sridhar 2006).

A further factor for the lack of competitiveness of CDMA was handset costs. Qualcomm fixed the royalty on CDMA handsets at about 7 per cent. This made them less affordable compared to GSM handsets. However, it has to be pointed out that even GSM handsets have some royalty attached to them. InterDigital,[1] a US high-tech corporation, retains a strong portfolio of patented GSM technologies that it licences to handset manufacturers worldwide. However, due to the large sale volumes of GSM handsets, the burden on the end user is much less compared to CDMA handsets. Unless Qualcomm reduced royalty fee, handset prices will not drop and the sales in a price sensitive market such as India could not increase. There were 35 dual technology licences issued across the 22 circles on the announcement of the dual licensing policy.

Flood of New Licences

The huge growth of mobile subscribers prompted many applicants, highly qualified and less so, to apply for the UAS licence. The speculation motive must be factored into the considerations of some applicants, given the perceived value and the low price of the licence. A total of 122 licences were issued across all circles. There were controversies related to a sudden advancing of the last date for applications, and discrepancies on deciding the first-come-first-serve criterion

---

[1] www.interdigital.com

Table 10.5   2008 policy outcomes

Parameters	March 2008	March 2011
Total number of players (most circles)	7	12
Subscribers (in millions)	261.1	846.0
HHI	1590	1394
Teledensity	25.0	70.9
Rural share in total	24.0	33.4

Source: Authors' own.

for assigning licences and associated spectrum. These controversies involved points of law, as well as issues related to the wisdom of subsidizing spectrum and letting in a large number of players.

Table 10.5 highlights some of the outcomes of the policy.

Fall in Price of Service

Table 10.5 clearly indicates the fall in HHI (refer to Glossary for definition), indicating hyper competition due to the introduction of more players. Through their intense competition, the telecom companies revived the value of the Indian paisa. A few of the new operators, introduced 'paisafication' of air time, offering even STD calls at the rate of a paisa per second.

An important concept that made this one paisa revolution possible is the 'price point pack' (PPP), an innovative marketing and packaging concept that was designed to sell a small portion or a single-use portion of the product at an affordable price. The first PPP ever introduced was some 60 years ago! Brooke Bond and Lipton, leading tea companies in India, launched what was called the 'paisa pack' which was a paper envelope that contained tea worth one paisa (then 1/64th of the Indian rupee) which could make 1 or 2 cups of the beverage. This model was followed diligently by Pan Masala and then later on by FMCG companies to sell shampoos in sachets. PPP has been successfully adopted by the Indian telecom operators to sell the perishable commodity often referred to as 'Erlangs' of network capacity (Sridhar and Sridhar 2010).

Significant Decline in ARPU

The subscriber base grew sharply with over 10 million new subscribers being added every month. The growth in subscriber base post 2008 along with key reasons for the growth of the market is shown in Figure 10.3.

The average revenue per user (ARPU) for both GSM as well as CDMA service providers continues to decline as shown in Figure 10.4. This was due to the sharp fall in call rates alluded to earlier as well as a reduction in usage as the mobile market brought lower income subscribers into its fold. Indian subscribers have the cheapest call rates in the world.

Of course, part of the subscriber growth may have been spurious with operators showing greater growth to qualify for their next tranche of spectrum. In order to identify fake subscribers, since April 2011, TRAI has switched to the practice of counting only those

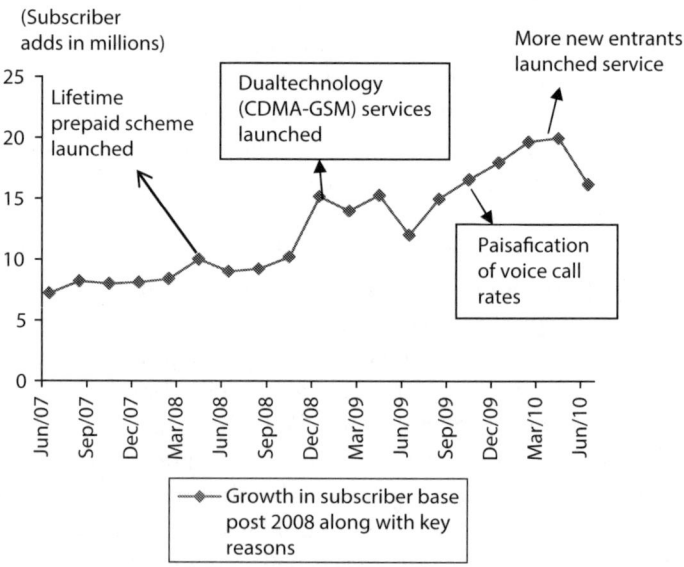

Figure 10.3   Growth in subscriber base post 2008 along with key reasons
Source: Authors' own.

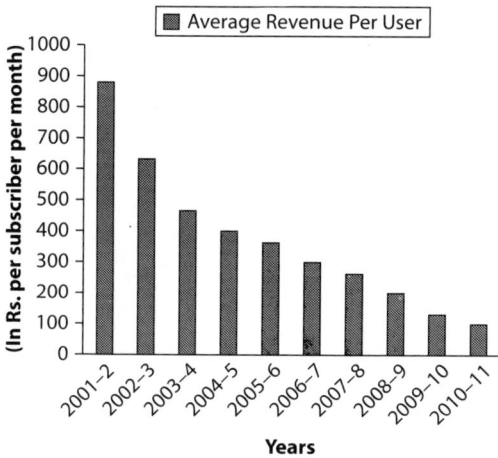

Figure 10.4   Decline in ARPUs post 2008

Source: Authors' own.

subscribers who feature in the visiting location register. The earlier practice of counting anyone who was listed in the home location register appears to have overstated subscribers by 20–25 per cent. To the extent that the spurious subscribers increased after the 2008 licensing, the actual rates of growth will be lower and the ARPUs higher.

Valuation: Sharp Rise and Steep Decline

Many new entrants were the beneficiaries of FDI at valuations that were many times in excess of the subsidized fee paid for the licence. For example, Norwegian telecom major Telenor acquired 67.5 per cent of Unitech Wireless for Rs. 6,120 crore, implying a valuation of Rs. 9,067 crore. The acquisition was made soon after Unitech acquired the licence, implying a 5.5 times appreciation. Other acquisitions exhibited similar growth multiples.

However, after a year, valuations of companies have shown a steep decline with the bellwether Airtel not being left behind in the race to the bottom (Table 10.6).

Table 10.6   Airtel valuation ratios

Key valuation ratios		2006–07	2007–08	2008–09	2009–10	2010–11
Enterprise value/ revenues	Times	8.1	6.1	3.5	3.0	3.2
Enterprise value/ EBITDA	Times	20.1	14.8	9.4	7.1	9.4
Market cap/ sales	Times	7.9	5.8	3.2	2.8	2.3
Enterprise value/ subscriber	Rs.	40,154	26,629	13,820	9,909	12,034

Source: Airtel (2013).

The Aftermath: Landmark Quashing of Licences by Supreme Court

The policy got mired in controversy due to irregularities in the process used for allocation of licences and spectrum. The two main issues involved were:

(1) Failure to revise the spectrum price that was pegged at the value determined in the fourth licence auction in 2001.

(2) The discretion used in fixing the last date for the receipt of applications for licences and in determining the order of the applicants for assigning spectrum. The last date for acceptance of applications for UAS licences was set at 1 October 2007. However, on 2 November 2007, DoT decided that only applications received up to 25 September 2007 would be processed. In total, 575 applications were received until 1 October 2007. The 343 applications made between 26 September 2007 and 1 October 2007 were made post hoc ineligible for immediate spectrum licence. The arbitrariness in the advance of the due date became one of the important issues for litigation (Swamy 2011).

There were widely varying estimates of the loss to the government's exchequer due to these irregularities. The Comptroller and Auditor General (CAG) of India estimated the loss to be up to Rs. 1.76 lakh crore based on the price discovered in the auction of 3G and BWA spectrum in 2010. However, the government continued to argue that lower one-time fee for spectrum increased competition and resulted in lowering of service prices making voice telephony affordable for almost all sections of the population. The government argued that the positive spillover effect of mobile penetration on economic development justified the low price and the high degree of competition.

All the speculations were put to rest on 2 February 2012 when the Supreme Court of India delivered its judgment on a public interest litigation (PIL) directly related to the '2G spectrum scam'. The Supreme Court declared the allotment of spectrum as 'unconstitutional and arbitrary' and quashed all the licences issued on or after 1 January 2008 during the tenure of A. Raja (then the Minister for Communications and IT), the main official accused in the 2G scam case (Singhvi & Ganguly 2012). The court further said that A. Raja 'wanted to favour some companies at the cost of the public exchequer' and 'virtually gifted away an important national asset'. It is to be noted that the Supreme Court order did not specify the number of licences quashed. The Supreme Court also set a deadline for assigning the spectrum vacated by the cancelled licensees.

Table 10.7 gives the list of firms and the associated number of licences amounting to 122 that were determined by DoT as cancelled.

There are controversies surrounding the determination of the quashed licences as well. For example, Tata Teleservices applied for a dual technology GSM licence and spectrum in 19 service areas and was given the letter of intent for its applications on 10 January 2008. However, these did not appear in the table released by DoT.

Liberalization of Spectrum 2008 Onwards

The tumultuous events following the 2008 licensing policy hastened the move to a liberalized spectrum regime. However, the first steps in this direction were taken not after 2008, but as early as 2006, when TRAI released its consultation on the auction of the 3G spectrum. The auction of the 3G spectrum was followed by the release

Table 10.7   Companies whose licences were cancelled

Name of company	LSA for which licences were cancelled	Number of licences cancelled
Uninor	Pan India	22
Sistema Shyam Tele Services Limited, now MTS India	Pan India except Rajasthan	21
Loop Mobile formerly BPL Mobile	Pan India except Mumbai	21
Videocon Telecommunications Limited	Pan India except Punjab	21
Etisalat-DB	In Andhra Pradesh, Bihar, Delhi, Gujarat, Haryana, Karnataka, Kerala, Madhya Pradesh, Maharashtra, Mumbai, Punjab, Rajasthan, Tamil Nadu (including Chennai), UP (East) and UP (West)	15
Idea Cellular & Spice	Andhra Pradesh, Assam, Delhi, Haryana, Jammu & Kashmir, Karnataka, Kolkata, Maharashtra, North East, Odisha, Punjab, Tamil Nadu (including Chennai), West Bengal	13
S Tel	Assam, Bihar, Himachal Pradesh, Jammu & Kashmir, North East and Odisha	6
Tata Teleservices	Jammu & Kashmir, Assam and North East	3
Total		**122**

Source: Authors' own.

of the New Telecommunications Policy of 2012 and the auction of spectrum obtained from quashed 2G licences. This was also the time when many licences were coming up for renewal. The formulation of a renewal policy in line with the new thinking is also one of the highlights of this phase.

Start of Wireless Broadband: 2010

TRAI released a consultation paper about 3G licences in June 2006 (TRAI, June 2006, clause 11.3.1. page 3) to usher in wireless broadband in the country. With sparse landline broadband deployment, wireless with its cost advantages seemed to be the more appropriate technology. However, due to various reasons, including delay in the release of spectrum by Defence, the 3G auctions were repeatedly postponed. Finally, in line with TRAI's recommendation, the spectrums for 3G and broadband wireless access (BWA) were assigned in 2010. The simultaneous ascending auction described in Chapter 4 was used for the allocation of spectrum. For the first time in the history of spectrum allocation in India, the licence was explicitly de-linked from spectrum. The information memorandum clearly stated the following (DoT October 2009):

> For the spectrum in 2.1GHz band ('3G Spectrum'):
> Any entity:
>
> (1) that holds a UAS/ Cellular Mobile Telephone Service ('CMTS') license; or
> (2) that:
>
>   (a) has previous experience of running 3G telecom services either directly or through a majority-owned subsidary; and
>   (b) gives an undertaking to obtain a UAS licence through a New Entrant Nominee UAS Licensee as per the DoT guidelines before starting telecom operations can bid for 3G Spectrum (subject to other provisions of the Notice).

The memorandum clearly implied that a licence was a prerequisite for providing service but not for obtaining spectrum.

The government auctioned spectrum blocks, each of 2×5 MHz in the 2.1 GHz band with specified reserve prices for different categories of LSAs. Note that 2 × 5 MHz is the minimum carrier requirement for providing 3G services using WCDMA technology in the 2.1 GHz band. A maximum of four blocks were auctioned in each LSA. One block was given to government operators (that is, BSNL and MTNL) one year in advance (that is, in 2009), subject to them paying the winning auction price. Due to the controversial assignment of spectrum in 2008 and the ensuing political upheaval, the incumbents were not assigned any 2G spectrum post 2008 based on the extant subscriber based criterion. With low visibility on the future availability of 1800 MHz for their 2G services, most of the incumbents approached the 3G auction to garner spectrum for enhancing their 2G services (that is, mainly voice services that continued to see rapid growth in the subscriber base). As a result auction prices were very high, provoking fears of a winners' curse.

The government also auctioned two blocks of 20 MHz for BWA service soon after the closure of the 3G spectrum auction. The idea behind auctioning large-sized blocks was that the time division duplexing technology which uses the entire spectrum band is spectrally efficient compared to frequency division duplexing. The WiMAX Forum has published three licensed spectrum profiles: 2.3 GHz, 2.5 GHz, and 3.5 GHz, in an effort to drive standardization and decrease costs. Interestingly, a 20 MHz spectrum block in the 2.5 GHz band was assigned to the government operators BSNL and MTNL across all service areas along with the 3G spectrum in 2009 itself. In fact, BSNL floated a tender for managed services of its WiMAX deployment subsequent to the allocation of spectrum. However, nothing much happened thereafter, and due to their inability to roll-out services, the debt-laden BSNL and MTNL plan to surrender their spectrum to get a waiver on spectrum charges. Thus far, India's position was to favour WiMAX as the 4G technology. However, as discussed in Chapter 1, WiMAX as a technology was dying due to low adoption by operators and LTE was gaining prominence as the 4G technology. The spectrum blocks auctioned were in the 2.3 GHz band, supported by both WiMAX and LTE.

There was one significant change in the eligibility criterion for the BWA spectrum auction. Apart from UAS and CMTS licence holders, ISPs were also allowed to participate in the auction process. The thinking behind this was that wireless was the technology to be nurtured to improve broadband penetration in the country because of escalating costs and declining subscriber interest in wire-line broadband. Hence, it was felt that the ISPs with their Internet backbone would be able to reach end consumers through wireless access networks. Moreover the roll-out conditions for the BWA service was much more focussed on rural areas compared to 3G services, indicating the government's focus on improving rural broadband connectivity using wireless access. Table 10.8 compares the features of the two auction designs.

However, as indicated in Chapter 1, the high-frequency 2.3 GHz is not well-suited for rural area coverage. Moreover, the ecosystem for supporting LTE in the unpaired frequency band in 2.3 GHz is still evolving. This uncertainty may have dampened bids for the BWA spectrum.

The financial results of the auction are given in Table 10.9. For 3G auctions, there were 183 rounds over 34 days, with about 5 rounds every day yielding a total revenue of Rs. 50,931 crore and an additional Rs. 16,751 from BSNL/MTNL.

For the BWA auction, there were 117 rounds over 16 days grossing Rs. 38,500 crore in total. Table 10.10 gives the complete results.

It is to be noted that though the initial reserve price for the BWA auction was 50 per cent of the 3G reserve price, the final bid prices varied widely between 30 per cent and just over 140 per cent of 3G prices. On average, the BWA prices were about 75 per cent of the 3G prices although the reserve price was only 50 per cent of the 3G reserve price. The relaxed eligibility criteria (that is, allowing ISPs to bid) in the BWA spectrum auction brought in new players such as Infotel, Qualcomm, Tikona, and Augere. The incumbent UAS licensees did not participate actively. One of the reasons could be that the BWA auctions were held just after the 3G spectrum auction.

As highlighted in Table 10.10, none of the operators, except the government ones (that is, BSNL and MTNL combined) acquired 3G spectrum for pan-India. This was also the case with the BWA spectrum, except for one private operator who won spectrum across India. As a result:

(1) The 3G spectrum winners, most of whom faced a critical shortage of the existing 2G spectrum, used the 3G spectrum initially

Table 10.8   The 3G/BWA auction process parameters

Parameter	Spectrum in 2.1 GHz (for 3G)	Spectrum in 2.3 GHz (for BWA)
Who is eligible?	Incumbent UAS or CMTS licensees; new entrant needs to acquire either of these licences	Incumbent UAS or CMTS licensees; category A or B type ISPs; new entrant needs to acquire one of the above licences
Block auctioned	Three/four blocks of 2 × 5 MHz depending on availability of spectrum in each LSA; one block given to government operators BSNL and MTNL	Two blocks of 20 MHz (unpaired) depending on availability of spectrum in each LSA; one block given to government operators BSNL and MTNL in 2.5 GHz
Reserve price for auction (in Rs. million)	Delhi and Mumbai: 3,200; Kolkata: 1,200; Category A: 3,200; Category B: 1,200; Category C: 300	Delhi and Mumbai: 1,600; Kolkata: 600; Category A: 1,600; Category B: 600; Category C: 150
Roll-out obligations	Metros: 90 per cent of the service area; street coverage in the relevant service areas within 5 years of the effective date	Metros: 90 per cent of the service area; street coverage in the relevant service areas within 5 years of the effective date
	Category A,B, and C: At least 50 per cent of the district headquarters (DHQs) out of which at least 15 per cent should be rural short distance charging areas, within 5 years of the effective date	Category A,B and, C: At least 50 per cent of the rural short distance charging areas, within 5 years of the effective date

Source: Authors' own.

Table 10.9 Financials of the 3G spectrum auction

LSA	Bid amount (in Rs. crore)	No. of winners	3G Winners	Total bid amount (in Rs. crore)
Andhra Pradesh	1,373.14	3	Bharti, Idea, Aircel	4,119.42
Assam	41.48	3	Reliance, Bharti, Aircel	124.44
Bihar	203.46	4	S Tel, Bharti, Reliance, Aircel	813.84
Delhi	3,316.93	3	Vodafone, Bharti, Reliance	9,950.79
Gujarat	1,076.06	3	Tata Tele, Idea, Vodafone	3,228.18
Haryana	222.58	3	Tata Tele, Idea, Vodafone	667.74
Himachal Pradesh	37.23	3	Bharti, S Tel, Idea, Reliance	111.69
Jammu & Kashmir	30.30	4	Idea, Aircel, Reliance, Bharti	121.20
Karnataka	1,579.91	3	Tata Tele, Aircel, Bharti	4,739.73
Kerala	312.48	3	Idea, Tata Tele, Aircel	937.44
Kolkata	544.26	3	Vodafone, Aircel, Reliance	1,632.78
Madhya Pradesh	258.36	3	Idea, Reliance, Tata Tele	775.08
Maharashtra	1,257.82	3	Tata Tele, Idea, Vodafone	3,773.46
Mumbai	3,247.07	3	Vodafone, Bharti, Reliance	9,741.21

(Continued)

Table 10.9 (Continued)

LSA	3G			
	Bid amount (in Rs. crore)	No. of winners	Winners	Total bid amount (in Rs. crore)
North East	42.30	3	Aircel, Bharti, Reliance	126.90
Odisha	96.98	3	S Tel, Aircel, Reliance	290.94
Punjab	322.01	4	Idea, Reliance, Tata Tele, Aircel	1,288.04
Rajasthan	321.03	3	Reliance, Bharti, Tata Tele	963.09
Tamil Nadu	1,464.94	3	Bharti, Vodafone, Aircel	4,394.82
Uttar Pradesh (E)	364.57	3	Aircel, Idea, Vodafone	1,093.71
Uttar Pradesh (W)	514.04	3	Bharti, Idea, Tata Tele	1,542.12
West Bengal	123.63	4	Bharti, Reliance, Vodafone, Aircel	494.52
Total	**16,750.58**			**50,931.14**
BSNL/ MTNL To Pay	16,750.58			16,750.58
Grand Total				**67,681.72**

Source: Authors' own.

Table 10.10 Financials of BWA spectrum auction

LSA	Bid amount (Rs. crore)	No. of winners	BWA Winners	Total bid amount (in Rs. crore)	BWA percentage of 3G bid
Andhra Pradesh	1,059.12	2	Aircel, Infotel	2,118.24	77.13%
Assam	33.02	2	Aircel, Infotel	66.04	79.60%
Bihar	99.28	2	Aircel, Infotel	198.56	48.80%
Delhi	2,241.02	2	Infotel, Qualcomm	4,482.04	67.56%
Gujarat	613.85	2	Infotel, Tikona	1,227.70	57.05%
Haryana	119.90	2	Infotel, Qualcomm	239.80	53.87%
Himachal Pradesh	20.66	2	Infotel, Tikona	41.32	55.49%
Jammu & Kashmir	21.27	2	Aircel, Infotel	42.54	70.20%
Karnataka	1,543.25	2	Bharti, Infotel	3,086.50	97.68%
Kerala	258.67	2	Infotel, Qualcomm	517.34	82.78%
Kolkata	523.20	2	Bharti, Infotel	1,046.40	96.13%
Madhya Pradesh	124.66	2	Infotel, Augere	249.32	48.25%
Maharashtra	915.64	2	Bharti, Infotel	1,831.28	72.80%

(Continued)

Table 10.10 (*Continued*)

LSA	Bid amount (Rs. crore)	No. of winners	BWA winners	Total bid amount (in Rs. crore)	BWA percentage of 3G bid
Mumbai	2,292.95	2	Infotel, Qualcomm	4,585.90	70.62%
North East	21.27	2	Aircel, Infotel	42.54	50.28%
Odisha	63.63	2	Aircel, Infotel	127.26	65.61%
Punjab	332.27	2	Bharti, Infotel	664.54	103.19%
Rajasthan	97.32	2	Infotel, Tikona	194.64	30.31%
Tamil Nadu	2,069.45	2	Aircel, Infotel	4,138.90	141.27%
Uttar Pradesh (E)	142.50	2	Infotel, Tikona	285.00	39.09%
Uttar Pradesh (W)	183.37	2	Infotel, Tikona	366.74	35.67%
West Bengal	70.97	2	Aircel, Infotel	141.94	57.41%
Total	**12,847.27**			**25,694.54**	**76.70%**
BSNL/ MTNL To Pay	**12,847.27**			**12,847.27**	
Grand Total				**38,541.81**	

Source: Authors' own.

for augmenting their 2G voice capacity. However, due to voice yielding very low ARPUs, they needed to invent innovative data packages to lure broadband customers, especially the youth who used pre-paid SIMs in their smartphones.

(2) Since none of the 3G spectrum winners had a pan-India presence, they resorted to inter-circle roaming arrangements to share spectrum across the service areas.

(3) The BWA spectrum holders adopted a wait-and-watch strategy as LTE equipment and handsets in 2300 MHz were still evolving. Fortunately for them, the roll-out obligation did not commence until 5 years after acquiring the licence. Meanwhile, the expectation was that prices of devices would come down due to large-scale adoption of the corresponding technology in China and elsewhere.

(4) The sole BWA spectrum holder across India, Infotel, sold its majority stake to Reliance Industries Ltd., which regained its presence in the telecom space after promoter Mukesh Ambani had to forego the group telecommunications arm, Reliance Communications to brother Anil in a family settlement.

(5) Some winners had never intended to provide services. For instance, Qualcomm, which basically did not have the intention of providing the service entered the race to promote its TD-LTE technology. In 2012, Airtel bought a majority stake in Qualcomm's operating arm that acquired BWA spectrum to augment its presence in the lucrative areas of Delhi, Mumbai, Haryana, and Kerala where it could not win spectrum in the auction.

New Telecom Policy, 2012

The uproar over the 2008 licensing policy brought home the need for a new framework of spectrum management. The new telecom policy was unveiled in 2012. In addition to a focus on increasing rural teledensity, promoting broadband for all, and using the domestic market to encourage indigenous manufacturing, the policy envisaged a liberalization of spectrum management. The key recommendations on spectrum were:

(1) To move at the earliest towards liberalization of spectrum to enable use of spectrum in any band to provide any service in

any technology as well as to permit spectrum pooling (referred to as 'leasing' in this book), sharing, and later trading to enable optimal utilization of spectrum through an appropriate regulatory framework.

(2) To undertake a periodic audit of spectrum utilization to ensure its efficient use.

(3) To re-farm spectrum and allot alternative frequency bands or media to service providers from time to time to make spectrum available for introduction of new technologies for telecom applications.

(4) To prepare a roadmap for availability of additional spectrum every 5 years.

(5) To make available adequate globally harmonized IMT spectrum in 450 MHz, 700 MHz, 1800 MHz, 1910 MHz, 2.1 GHz, 2.3 GHz, 2.5 GHz, and 3.5 GHz bands and other bands to be identified by ITU for commercial mobile services.

(6) To identify additional frequency bands periodically, for exempting them from licensing requirements for operation of low power devices for public use.

(7) To consider requirement of spectrum in certain frequency bands in small chunks at specified locations for encouraging indigenous development of technologies/products and their deployment.

(8) To review the existing geographical unit of allocation of spectrum with a view to identifying scope for optimization.

(9) To promote use of white spaces with low-power devices, without causing harmful interference to the licensed applications in specific frequency bands by deployment of software defined radios (SDRs), and cognitive radios (CRs), etc.

(10) To establish and strengthen the Institute of Advanced Radio Spectrum Engineering and Management Studies (IARSEMS) as a government society for undertaking policy research in radio spectrum engineering, management/radio monitoring and related aspects.

A more comprehensive licensing framework with a universal licence, including provision of infrastructure services, operation of a virtual network in the MVNO mode, and provision of all services to the end customer, was also proposed.

Legacy Issues in Liberalization

The following legacy issues had to be addressed while the liberalization regime was being introduced:

Migration of Licences

The spectrums that were given to the operators from 1995 to 2007, through a combination of administrative and market processes are technology and service specific (that is, 2G services for GSM and CDMA technologies). These needed to be migrated to technology and service agnostic forms. The issue of creating a level playing field with the 2008 licensees who have to pay a market determined fee also needed to be addressed.

After much prolonged back and forth between various government bodies, the Empowered Group of Ministers (EGoM) decided in late 2012 to collect a one-time spectrum fee from all operators with more than 4.4 MHz (2.5 MHz in case of CDMA operators) for the remaining period of their licences. But the group added a new twist by deciding to collect an additional levy from operators that were given more than 6.2 MHz spectrum (5 MHz in case of CDMA operators) July 2008 onwards. This proposal was in contrast to the earlier recommendation by the Attorney General to levy one-time fee retrospectively from 2008 for spectrum holding beyond 6.2 MHz for GSM incumbent operators (5 MHz for CDMA operators). This one-time fee retrospectively from 2008 is to provide a level playing field between incumbents who were assigned spectrum using administrative procedure and the new entrants who got it through auction. However, this approach fails to recognize the heterogeneity of the spectrum allocation procedure for different licence holders. The licensees can be divided into those who got spectrum in 1995, in 2001, and post 2001.

In 1995, there was a 'Beauty Contest' with a fixed fee for the four metros and a single-stage sealed bid auction for the remaining 18 service areas. However, since the auction resulted in a 'winner's curse' with the operators struggling to pay the yearly bid amount, the government bailed them out in 1999 by shifting them to a 'revenue sharing scheme' with an upfront fee equivalent to the pay-out for 2.9 years to cover the 1995–99 time period.

In 2001, GSM operators got bundled spectrum of 4.4. MHz in a simultaneous ascending auction for the fourth operator licences. All additional allocations for these operators were allocated administratively thereafter.

Third operators including GSM and CDMA operators and dual technology operators, who got spectrum post 2001 and were charged a fixed fee equivalent to the 2001 auction determined price using the administrative pricing mechanism.

In all these cases there were parts of spectrum which had been assigned and priced administratively (DoT 2009, p. 20). Given this heterogeneity, it is simpler and fairer to charge all the spectrum holding of the operators through market determined price and not to fix the threshold of 4.4 MHz or 6.2 MHz. As stated in the Subodh Kumar Committee report:

> The government need not exercise itself unduly about ensuring an absolute level playing field between licensees who entered the market at different points in time. Variable pricing of resources for entrants at different times happens with other natural resources (e.g., land for industrial development) as well. Early or late entry, each comes with a set of advantages and disadvantages, and is part of the business proposition. The government needs to consider only whether the objectives of fair competition, increasing teledensity, and efficient spectrum use are met while charging appropriate licence fees.

The principle of the level playing field should be strictly implemented from the time of the decision to end the administrative regime. The cut-off date for the administrative regime as per the Supreme Court verdict was 2 February 2012. Hence we recommended that all spectrum holding should be priced prospectively until the end of the respective licence period using current market determined price.

We recommended a prospective charge in all cases because levying charges retrospectively is generally a bad practice. The spectrum charges paid by the operators as a percentage of the AGR accumulating to a total of more than Rs. 1,200 crore per quarter, partly cover the opportunity cost of spectrum in the administrative pricing regime. Moreover, the additional spectrum allocation in the range of 1.0–1.2 MHz in case of GSM and 1.25 MHz in the case of CDMA based on the 'subscriber based criteria' restricted the use of spectrum to certain types of services. Hence, retrospective pricing based on auctions in a liberalized regime cannot be supported.

This proposal places all operators in a level playing field. All pay prospectively for airwaves given through the non-auction method, equivalent to an auction determined price, and get to use the liberalized spectrum for any technology/service.

Renewal of Licences

The spectrum in the 900 MHz band assigned to operators (Bharti Airtel, Vodafone, and Loop Mobile) in the metro areas of Delhi, Mumbai, and Kolkata in 1995 is nearing its tenure of 20 years and is coming up for extension in November 2014. Subsequent assignments in this band are coming up for renewal shortly thereafter. The 800 MHz spectrum that was also administratively assigned during 1997–8 is also coming up for renewal starting in 2017. Table 10.11 gives the expiry timetable of the 900 MHz and 800 MHz spectrum.

The possible approaches that can be taken to achieve the objective of migrating to advanced technologies while maintaining the continuity of services have been discussed in Chapter 6. To recall, using 900 MHz as an example, the different approaches are:

(1) Complete liberalization: Take back all 900 MHz spectrum at the end of licence period and place all available 1800 MHz spectrum and 900 MHz spectrum on auction for all licence holders. Let auctions decide spectrum assignments.

(2) Complete continuity: Let operators retain all their spectrum, provided they pay the market price for it. Auction available 900 MHz spectrum along with other spectrums.

Table 10.11 Timeline for the expiry of licence and associated spectrum

Year	2014	2015	2016	2017	2018	2020	2021	2024
800 MHz	0	2	0	7	1	20	22	13
900 MHz	7	25	4	2	1	20	1	11

Source: TRAI (2012).

(3) Middle ground: All incumbents to keep a part of their spectrum or are given lower technology spectrum (1800 MHz in this case) in lieu of part or whole of the 900 MHz spectrum. Levy market prices on assured spectrum. The rest of the spectrum holding should be surrendered to the government for refarming and auctioning.

  (a) Allow incumbents to keep some amount of 900 MHz spectrum (example, 5 MHz) so that they can continue basic operations.

  (b) Provide 1800 MHz spectrum in lieu of a part of the 900 MHz spectrum (example, 5 MHz) so that operators can continue basic operations.

  (c) Replace all 900 MHz spectrum with 1800 MHz spectrum.

The government adopted approach 3(c). Its decision was in line with the TRAI recommendations (TRAI April 2012, clause 2.82, p. 24):

The Authority therefore recommends that the refarming of spectrum in the 800 MHz and 900 MHz bands should be carried out progressively at an early date but not later than the due date of renewal of the licences. The spectrum available with the service providers in the 900 MHz band should be replaced by spectrum in the 1800 MHz band, which should be charged at the price prevalent at the time of refarming. The Authority also recommends that the Government must actively explore the possibility of refarming of the spectrum in the 900 MHz band immediately, by invoking the authority to change the licence conditions.

However, incumbents were protesting that this is an unfair bargain as 900 MHz has far better propagation characteristics than 1800 MHz. Moreover, as per licence terms, the first right of refusal is with the incumbents for continuity of existing operations. Further, the reserve price of 900 MHz was fixed at twice that of 1800 MHz and hence it would become expensive for the incumbents to retain their 900 MHz holdings.

The controversial SBC for spectrum allocation was finally terminated but ascending spectrum usage charge continued to be in place. The incompatibility of the SBC with the liberalization of spectrum has been explored in detail in chapters 6 and 7.

Table 10.12 Spectrum usage charges

Spectrum slab	Percentage of AGR
GSM	
Up to 4.4 MHz	3
Up to 6.2 MHz	4
Up to 8.2 MHz	5
Up to 10.2 MHz	6
Up to 12.2 MHz	7
Up to 15.2 MHz	8
CDMA	
Up to 5 MHz	3
Up to 6.25 MHz	4
Up to 7.5 MHz	5
Up to 12.5 MHz	7
Up to 15 MHz	8

Source: TRAI (2013).

Table 10.12 gives the annual spectrum charge amount even after-migrating to a market-oriented method of assigning spectrum. This certainly is a double whammy for the operators.

The November 2012 Auctions of Spectrum from Quashed Licences

As per the directions of the Supreme Court order dated 2 January 2012, a consequential reference was made by DoT to TRAI. TRAI submitted its recommendations dated 23 April 2012 on the 'Auction of Spectrum' (TRAI 2012). Based on this and the recommendations of the Telecom Commission, the EGoM considered various issues including the method of allocation of spectrum, spectrum pricing, and quantum of spectrum to be released for mobile services. The auction of 1800 MHz was the first step in this direction.

In order to cater to the demands of both the incumbents for acquiring additional spectrum and the new entrants (whose licences were quashed by the Supreme Court verdict), the government initi-ated steps to auction spectrum. TRAI in its recommendations (TRAI

2012) worked out the amount of spectrum available for auction in 1800 MHz after reserving the required amount for the refarming of the 900 MHz spectrum as explained in the previous section.

Auction Design

The following were the assignment rules laid down for the 1800 MHz band (DoT 2012):

(1) Block size shall be 1.25 MHz (paired); a minimum of 8 blocks each of 1.25 MHz (totalling 10 MHz) were put to auction in all service areas. In addition, it was announced that a provision will be made for spectrum up to 3 blocks each of 1.25 MHz (3.75 MHz), wherever available (after reserving spectrum for refarming of the 900 MHz band, for the licences expiring during 2014 to 2016 and also taking care of guard band requirements), for topping up the 8 blocks of spectrum put to auction. Hence, up to a total of 11 blocks each of 1.25 MHz was put up for auction to meet the requirements of new entrants.

(2) Existing operators were allowed to take maximum 2 blocks of 1.25 MHz each in each service area.

(3) New entrants were required to bid for minimum of 4 blocks each of 1.25 MHz in each service area.

(4) New entrants were also allowed to bid for 1 additional block of 1.25 MHz in each service area.

The following were the assignment rules for the 800 MHz band (DoT 2012):

(1) Block size shall be of 1.25 MHz (paired); 3 blocks each of 1.25 MHz (3.75 MHz) be put to auction. In addition, a provision be made for spectrum of one block of 1.25 MHz, wherever available, for topping up the 3 blocks of spectrum put to auction, to meet the requirements of new entrants.

(2) Existing operators will be allowed to bid for one block of 1.25 MHz in each service area.

(3) A bidder who is a new entrant in a particular service area will be required to bid for a minimum of 2 blocks each of 1.25 MHz in each service area.

(4) The new entrant bidder will be allowed to bid for one additional block of 1.25 MHz in each service area.

Table 10.13 gives the amount of 1800 MHz and 800 MHz spectrum made available for auction.

While the government had to obey the Supreme Court's orders in auctioning the spectrum, it also had to be seen as not giving away spectrum cheap. Despite many petitions by industry the reserve price for 1800 MHz continued to be close to the price discovered for the 3G auction in 2010.

The reserve price in any auction is based on an estimate of the value of the object being auctioned and the discount applied in order to give sufficient room for the market to arrive at the true value through the bidding process without incurring the risk of collusion (Klemperer 2002). In TRAI's (2012) recommendation, the estimate of the value of 1800 MHz draws on the accompanying recommendation to liberalize the use of spectrum, allowing any band of spectrum to be used with any technology. From this policy stance, and from the availability of LTE devices in the 1800 MHz band, TRAI concluded in 2012 that the relative value of 1800 MHz and 2100 MHz spectrum was linked only to their relative propagation capability, that is, their 'intrinsic value', thereby ruling out factors related to the 'extrinsic value' of spectrum (Alden 2012; Mölleryd and Markendahl 2012) such as the state of the ecosystem associated with various spectrum bands. As per this reasoning, the 1800 MHz band has 1.2 times the propagation characteristics of the 2100 MHz band and so is 1.2 times more valuable. The value of 2100 MHz is obtained from the auction of 2010. TRAI further applied an average prime lending rate of 12.63 per cent to the 2010–11 price to arrive at an estimate of the value of the 1800 MHz spectrum.

Next, the discounting factor was applied. In arriving at this factor the TRAI recommendation argued, 'a study of various auctions held globally in last 3–4 years reveals that the reserve prices are generally around 0.5 times the final prices. However, in the context of Indian telecom sector, where the demand for spectrum is considerably higher, the Authority has decided to use a factor of 0.8 to determine the reserve price.' Thus as per TRAI:

Reserve price of 1800 MHz = Price of 2100 MHz × 1.2 × (1.1263)² × 0.8

Table 10.13 Availability of spectrum for auction in the 1800 MHz and 800 MHz bands

S.No.	Service area	1800 MHz		800 MHz		
		A	B	C	D	E
1	Delhi	8	Nil	1	1	Nil
2	Mumbai	8	Nil	1	1	Nil
3	Kolkata	8	3	1	1	Nil
4	Maharashtra	8	2	1	1	Nil
5	Gujarat	8	Nil	1	1	1
6	Andhra Pradesh	8	3	1	Nil	1
7	Karnataka	8	3	1	1	1
8	Tamil Nadu	8	3	1	1	1
9	Kerala	8	3	1	1	1
10	Punjab	8	Nil	1	Nil	Nil
11	Haryana	8	2	1	1	1
12	Uttar Pradesh (West)	8	2	1	1	1
13	Uttar Pradesh (East)	8	Nil	1	1	1
14	Rajasthan	8	Nil	Nil	Nil	Nil
15	Madhya Pradesh	8	3	1	1	1
16	West Bengal	8	3	1	1	1
17	Himachal Pradesh	8	Nil	1	1	1
18	Bihar	8	2	1	1	1
19	Odisha	8	3	1	1	1
20	Assam	8	3	1	1	1
21	North East	8	3	1	1	1
22	Jammu & Kashmir	8	3	1	1	1

Source: DoT (2012).

Note: A: Number of 1.25 MHz block (paired), minimum of 8 as specified in (i) of the 1800 MHz guidelines; B: Number of 1.25 MHz (paired) blocks, maximum up to 3 wherever available after reserving spectrum for refarming against allocation in 900 MHz band for the licences expiring during 2014 to 2016, as per clause (i) of the 1800 MHz guidelines; C: Number of blocks of 2.50 MHz (paired) (first and second blocks of 1.25 MHz); D: Number of 1.25 MHz block (paired) (third block); C and D as per clause (i) of 800 MHz guidelines; E: Number of 1.25 MHz block (paired) wherever available (fourth block), as per clause (i) of 800 MHz guidelines.

The reserve prices of 800 MHz and 900 MHz spectrum were fixed at *twice* the reserve price of 1800 MHz spectrum and the reserve price of 700 MHz at *four* times the reserve price of 1800 MHz, based on their relative propagation capabilities. Table 10.14 gives the reserve prices so computed for various bands (TRAI 2012). For 2100 MHz, the winning auction price for 3G spectrum during 2010 was adjusted for a one-year interest rate. Similarly, for 2300 MHz, the winning auction price for BWA spectrum during 2010 was adjusted for a one-year interest rate. The final set of reserve prices per megahertz recommended by TRAI are given in Table 10.14.

It is clear that the reserve price set for 1800 MHz spectrum will determine the reserve prices in all auctions and will play an important role in determining the cash outflows of different players (especially owing to the choice of a high discounting factor of 0.8). Therefore, the choice of the reserve price and the overall auction design for 1800 MHz acquires immense importance for the future of India's telecom industry.

The benchmarking of the value of the 1800 MHz spectrum to the price discovered in the auction for 2100 MHz spectrum was problematic for a number of reasons. As shown in the TRAI recommendation (TRAI 2012), at present, 3G devices in the 1800 MHz band are far fewer in number than in the 2100 MHz band. In fact, the 1800 MHz is more adopted for deploying LTE technologies than 3G technology such as WCDMA. Further, voice services will continue to dominate for the next several years and it is plausible that the 1800 MHz band will continue to be used mainly with voice services in the foreseeable future. Finally, while the new blocks available for auction are contiguous, the 1800 MHz band presently held by operators (for which they will be charged based on the auction prices), suffers from fragmented assignment, compared to that of 2100 MHz band. Their assignment is in chunks of a 200 KHz discontinuous spectrum, whereas the assignments in the case of the 2100 MHz band are in chunks of 5 MHz. A fragmented band with a nascent 3G ecosystem cannot be treated at par with a harmonized band which has been used for 3G services for a number of years.

Economic models for the valuation of 1800 MHz show that the value of 1800 MHz on a pan-India basis is about 60 per cent of the value of the 2100 MHz. Abstracting away from scale effects,

Table 10.14   Reserve prices for the various bands recommended by TRAI

S. No.	Licence service area	700 MHz	800 MHz	900 MHz	1800 MHz	2100 MHz	2300 MHz
1	Delhi	2869.04	1434.52	1434.52	717.26	747.17	126.20
2	Mumbai	2808.56	1404.28	1404.28	702.14	731.43	129.13
3	Kolkata	470.76	235.38	235.38	117.69	122.60	29.46
4	Maharashtra	1087.96	543.98	543.98	271.99	283.34	51.56
5	Gujarat	930.76	465.38	465.38	232.69	242.39	34.57
6	Andhra Pradesh	1187.72	593.86	593.86	296.93	309.31	59.64
7	Karnataka	1366.56	683.28	683.28	341.64	355.89	86.91
8	Tamil Nadu	1267.12	633.56	633.56	316.78	329.99	116.54
9	Kerala	270.32	135.16	135.16	67.58	70.39	14.57
10	Punjab	278.52	139.26	139.26	69.63	72.54	18.71
11	Haryana	192.56	96.28	96.28	48.14	50.14	6.75
12	Uttar Pradesh (West)	444.64	222.32	222.32	111.16	115.79	10.35
13	Uttar Pradesh (East)	315.32	157.66	157.66	78.83	82.12	8.02
14	Rajasthan	277.68	138.84	138.84	69.42	72.32	5.48
15	Madhya Pradesh	223.48	111.74	111.74	55.87	58.20	7.02
16	West Bengal	106.96	53.48	53.48	26.74	27.85	4.00
17	Himachal Pradesh	32.2	16.10	16.10	8.05	8.39	1.16
18	Bihar	175.96	87.98	87.98	43.99	45.83	5.59
19	Odisha	83.92	41.96	41.96	20.98	21.85	3.58
20	Assam	35.88	17.94	17.94	8.97	9.34	1.86
21	North East	36.6	18.30	18.30	9.15	9.53	1.20
22	Jammu & Kashmir	26.2	13.10	13.10	6.55	6.83	1.20

Source: TRAI (2012).

there is also a difference in the distribution of value across circles. 3G spectrum is more valuable in metros and category A circles while the 2G spectrum is more valuable in category B and C circles. None of these nuances were taken into account.

Due to the uproar among telcos, the government at the time of the auction shaded the reserve prices of 1800 MHz by about 25 per cent. The reserve price of the 800 MHz spectrum was reduced from twice that of 1800 MHz to 1.3 times. The final set of reserve process for the 1800 MHz and 800 MHz that were put up for auction are given in Table 10.15 (DoT 2012). This translated to about Rs. 14,000 crore for a pan-India spectrum of 5 MHz in the 1800 MHz band.

Results of the Auction

Given the fallacies in selecting the reserve price the outcome of the auction was self-evident. The simultaneous ascending auction that took place in 2012, concluded on 14 November 2012 in just 2 days and consisted of 14 rounds (compared to the 34 days and 183 rounds of the 3G spectrum auction).

The government received bids worth Rs. 9,336 crore, far lower than its target of Rs. 28,000 crore for the 1800 MHz spectrum. None of the bidders bid for a pan-India spectrum for the 5 MHz block in 1800 MHz for which the reserve price was set at Rs. 14,000 crore on a pan-India basis. Out of the 144 blocks (each of 1.25 MHz) of spectrum on offer, 99 received bids. Delhi, Mumbai, Karnataka, and Rajasthan circles did not receive any bids. Further, there was no participation in the 800 MHz band in any of the service areas. The final results of the auction are given in Table 10.16.

In addition to the option of payment of the winning bid amount as a full upfront payment within 10 days of final results being declared, the following deferred option was also proposed for both the auctions held in November 2012 and March 2013 (DoT 2012):

(1) An upfront payment of 33 per cent in the case of the 1800 MHz spectrum and 25 per cent in the case of the 800 MHz spectrum of the final bid amount of one time charges for the spectrum be made within 10 days of the declaration of a successful bidder and final price;

Table 10.15 Reserve prices for various spectrum blocks auctioned during 2012–13

Service area	Reserve prices (in Rs. crore) per block of 1.25 MHz fixed for November 2012 auction			Reserve prices (in Rs. crore) per block of 1.25 MHz fixed for March 2013 auction		
	1800 MHz	*800 MHz*	*900 MHz*	*1800 MHz*	*800 MHz*	*900 MHz*
Delhi	693.06	900.98	1,386.12	485.14	450.49	970.28
Kolkata	113.72	147.84	227.44	79.60	73.92	159.21
Mumbai	678.45	881.99	1,356.90	474.92	440.99	949.83
Andhra Pradesh	286.91	372.98	573.82		186.49	0.00
Gujarat	224.84	292.29	449.68		146.15	0.00
Karnataka	330.12	429.16	660.24		214.58	0.00
Maharashtra	262.81	341.65	525.62		170.83	0.00
Tamil Nadu, Chennai	306.09	397.92	612.18		198.96	0.00
Haryana	46.52	60.48	93.04		30.24	0.00
Kerala	65.30	84.89	130.60		42.45	0.00
Madhya Pradesh	53.99	70.19	107.98		35.09	0.00
Punjab	67.28	87.46	134.56		43.73	0.00
Rajasthan	67.08	87.20	134.16	46.96	43.60	93.91
Uttar Pradesh (E)	76.17	99.02	152.34		49.51	0.00
Uttar Pradesh (W)	107.41	139.63	214.82		69.82	0.00
West Bengal, Andaman & Nicobar	25.84	33.59	51.68		16.80	0.00
Assam	8.67	11.27	17.34		5.64	0.00
Bihar	42.51	55.26	85.02		27.63	0.00
Himachal Pradesh	7.78	10.11	15.56		5.06	0.00
Jammu & Kashmir	6.33	8.23	12.66		4.11	0.00
North East	8.84	11.49	17.68		5.75	0.00
Odisha	20.27	26.35	40.54		13.18	0.00
Total	**3,499.99**	**4,549.99**	**6,999.98**		**2,274.99**	**2,173.23**

Source: DoT (2012).

Table 10.16 Results of the 1800 MHz spectrum auction held in November 2012

LSA	Winning price (in Rs. crore) per block of 1.25 MHz	Winners (no. of blocks won)	Total no. of blocks assigned	Total spectrum assigned (pair of MHz)	Total bid amount (in Rs. crore)	Percentage deviation from reserve price
Delhi		0	0	0	0.00	
Mumbai		0	0	0	0.00	
Kolkata	113.72	Idea (4)	4	5	454.88	0.00
Maharashtra	262.81	Telewings (4); Vodafone (1)	5	6	1,314.05	0.00
Gujarat	224.84	Telewings (4); Videocon (4)	8	10	1,798.72	0.00
Andhra Pradesh	286.91	Telewings (4)	4	5	1,147.64	0.00
Karnataka		0	0	0	0.00	
Tamil Nadu	306.09	Idea (4)	4	5	1,224.36	0.00
Kerala	65.30	Vodafone (1)	1	1	65.30	0.00
Punjab	67.28	Vodafone (1)	1	1	67.28	0.00
Haryana	46.52	Videocon (4); Vodafone (2)	6	8	279.12	0.00
Uttar Pradesh (W)	107.41	Videocon (4); Telewings (4)	8	10	859.28	0.00%

(Continued)

Table 10.16 (Continued)

LSA	Winning price (in Rs. crore) per block of 1.25 MHz	Winners (no. of blocks won)	Total no. of blocks assigned	Total spectrum assigned (pair of MHz)	Total bid amount (in Rs. crore)	Percentage deviation from reserve price
Uttar Pradesh (E)	76.17	Vodafone (1); Telewings (4); Videocon (4)	9	11	685.53	0.00
Rajasthan		0	0	0	0.00	
Madhya Pradesh	53.99	Videocon (4); Vodafone (2)	6	8	323.94	0.00
West Bengal	25.84	Idea (4); Vodafone (2)	6	8	323.94	0.00
Himachal Pradesh	7.78	Vodafone (1)	1	1	7.78	0.00
Bihar	46.43	Idea (1); Vodafone (2); Videocon (4); Telewings (4)	11	14	510.73	9.22
Odisha	20.27	Idea (4); Vodafone (2)	6	8	121.62	0.00
Assam	8.67	Bharti Airtel (1); Idea (4); Vodafone (2)	7	9	60.69	0.00
North East	8.84	Idea (4); Vodafone (2)	6	8	53.04	0.00
Jammu & Kashmir	6.33	Vodafone (2); Idea (4)	6	8	37.98	0.00
Total			**99**	**124**	**9,336**	

Source: Authors' own.

(2) There shall be a moratorium of 2 years for payment of the balance amount of one time charges for the spectrum, which shall be recovered in 10 equal annual instalments. The amount payable annually by the successful bidders shall include interest on such instalments for the period it remains deferred, while ensuring that the NPV of the bid amount is safeguarded by applying suitable interest rate.

Further, the panel of ministers on spectrum, headed by Finance Minister P. Chidambaram, decided that mobile operators whose permits were quashed, but who had taken part in spectrum auctions to secure new licences could avail the provision to set-off the entry fee they had paid in 2008. This resulted in Telewings, Videocon, Idea Cellular, and Sistema adjusting the fee that they had paid in 2008 against charges for the airwaves that they had won in the recent auctions. The result was that the government did not get any revenue at least for the first two years.

The following conclusions emerge from the results of the auction:

(1) Only two of the firms whose licences were quashed were active in winning spectrum—Telewings (majority owned by Telenor of Norway which ended its controversial partnership with Unitech Wireless and partnered with Sun Pharma to float this new company) and Videocon.
(2) Only Vodafone and Idea participated in the auction to acquire additional spectrum to supplement their existing holdings. Bharti Airtel participated in just one circle.
(3) Only Idea acquired 5 MHz in the services areas in which it did not have spectrum in order to attain a pan-India status.
(4) The final winning prices in all circles except Bihar were the same as the reserve prices indicating a possible fallacy in the setting of the reserve prices as explained earlier. This was despite the fact that the reserve prices mentioned were lower than the TRAI recommendation on account of downward revisions done by the government.

It was significant that no bids were received for any of the circles for the 800 MHz. The bidding took very little time to complete much to the agony of the government which was expecting to garner a large revenue to plug the fiscal deficit. However, the government and the

Communications and IT minister were also relieved as the results were seen by some as posing a question mark on the estimation of the revenue loss by the CAG.

However, it could be argued that the results of this auction do not prove that the spectrum was correctly priced in 2008. The total per Mega Hertz price in the circles that saw sales was close to 10 times the price in 2001. The reserve price in the circles where blocks were unsold was over ten times the 2001 price. Hence, the substantive point of the CAG—that of pricing 1800 MHz spectrum in 2008 at 2001 prices was inappropriate, is borne out by the 2012 auction. The liberalization of the 1800 MHz block in 2012 (in 2008 spectrum was still managed in the command and control framework) cannot be the reason for the higher prices given the uncertainty over the pace of development of 3G technology in 1800 MHz, and the likely use of this band for voice services.

Apart from the reserve price, another reason for the failure of the auction was that the government was planning to charge incumbents for spectrum beyond 4.4 MHz on the basis of the price discovered in the auction. Further, the plan for refarming, envisaged taking away 900 MHz for use with 3G/4G technologies and giving back an equal number of megahertz of 1800 MHz to be charged in line with the auction prices. Both these factors made incumbents stay away from the auction. The blocks where reserve prices were reasonable, that is, category B and C circles, were taken up by the incumbents to top-up their spectrum for voice services. In the metros and category A circles where data services are in demand, the value for the 1800 MHz spectrum blocks was perceived to be much lower than the reserve price fixed based on winning prices for data-oriented 3G spectrum.

This also indicates that only a suitably designed auction (that is, one with appropriate reserve prices) can determine the true market value. Despite the auctioning of the spectrum, annual spectrum charges varying from 3 to 8 per cent depending on spectrum holding continue to exist thus suppressing the industry with double taxation.

Re-auction in February 2013

The dismal failure of the November 2012 auction prompted the government to critically look at the reserve price, especially that of

800 MHz, which was pegged at about 2 times that of 1800 MHz, and for which there were no takers during the auction in November 2012. The government arbitrarily reduced the reserve price of 1800 MHz band by 30 per cent (for allocation in the circles in which it was unsold during the November auction namely, Delhi, Mumbai, Karnataka, and Rajasthan) and by 50 per cent for the 800 MHz band.

The government simultaneously put on auction airwaves in Delhi, Mumbai, and Kolkata for the 900 MHz band, since this spectrum block was assigned to the operators in 1995 and was nearing its licence tenure of 20 years.

The conditions for bidding for 800 MHz were almost the same as in November 2012. Each auctioned block was of size 1.25 MHz. Incumbents were allowed to bid for a maximum of one block while the new entrants were allowed to bid for a minimum of two blocks. The auction was completed after three rounds on 11 March 2013. There were no applicants for 1800 MHz and 900 MHz bands. For the 800 MHz auction, there was only one bidder—Sistema Shyam, a joint venture between Shyam Telecom and Sistema of Russia, whose licences were quashed earlier. The reserve prices and the results of the auction are given in Table 10.17.

As can be seen, only some of the circles were picked up, and that too by only one firm—Sistema Shyam, that was actively providing CDMA-based services across all circles before the quashing of the licences. Winning bid prices were equal to the reserve prices, indicating the possibility of inadequate participation of firms and, once again, a failure of markets. There were a number of circles where one block of spectrum remained unsold. There were also concerns raised by the Comptroller and Auditor General that this non-participation in the auctions during November 2012 and March 2013 might be the result of collusion between the operators to force the government to reduce reserve prices in future auctions. A further dampener for the auction was that the incumbent operators had moved the Delhi High Court seeking its intervention to stop the auction of the 900 MHz bandwidth, after they failed to get a response from DoT on their plea for renewal of licences for the 900 MHz spectrum band which they had bought.

Details of spectrum allotted to holders of quashed licences, spectrum put up for auction, and spectrum sold in different bands are

Table 10.17 Results of the 800 MHz spectrum auction held in March 2013

LSA	Winning price (in Rs. crore) per block of 1.25 MHz	Winners (no. of blocks won)	Total no. of blocks assigned	Total spectrum assigned (pair of MHz)	Total bid amount (in Rs. crore)	Reserve price (in Rs. Cr)	Percentage deviation from reserve price
Delhi	450.49	Sistema (3)	3	3.75	1,351.47		
Mumbai							
Kolkata	73.92	Sistema (3)	3	3.75	221.76	73.92	0.00
Maharashtra							
Gujarat	146.50	Sistema (3)	3	3.75	439.50	146.50	0.00
Andhra Pradesh							
Karnataka	214.58	Sistema (3)	3	3.75	643.74	214.58	
Tamil Nadu	198.96	Sistema (3)	3	3.75	596.88	198.96	0.00
Kerala	42.45	Sistema (3)	3	3.75	127.35	42.45	0.00
Punjab							
Haryana							

Uttar Pradesh (W)	69.82	Sistema (3)	3	3.75	209.46	69.82	0.00
Uttar Pradesh (E)							
Rajasthan							
Madhya Pradesh							
West Bengal	16.79	Sistema (3)	3	3.75	50.37	16.79	
Himachal Pradesh							
Bihar							
Odisha							
Assam							
North East							
Jammu & Kashmir							
Total			24	30	3,640.53		

Source: Authors' own.

given in Table 10.18. LSA-wise details of spectrum assignment are given in Table 10.19.

In a hearing held on 15 February 2013, Supreme Court directed: 'The entire spectrum released as a result of quashing of the licences on 2 February 2012 should be auctioned without further delay.' In order to comply with the Supreme Court order, DoT referred the matter to EGoM for fixing the reserve price and quantum of spectrum to be auctioned. The EGoM further referred the matter to TRAI for a recommendation during early July 2013. TRAI was given a maximum of 60 days to come up with its recommendations which would then be considered by the Telecom Commission and EGoM.

Meanwhile, the holders of the 900 MHz spectrum whose licences are expiring in November 2014 filed writ petitions in the Delhi High Court regarding their contractual rights for extending their UASL which at the time of issue was bundled with the 900 MHz spectrum. However, the petitions were disposed and they were asked to present the case before Secretary, DoT. In DoT's view, the applicants' cases did not hold water since, as per NTP 2012, spectrum was delinked from the licence. This necessitated that the applicants acquire any spectrum at market determined prices. Realizing this, one of the applicants, Vodafone, indicated that it was willing to pay Rs. 4,000 crore for the 900 MHz (Rs. 1,700 crore each in Delhi and Mumbai and Rs. 600 crore for Kolkata). This was much less than the reserve prices fixed by the government during the March 2013 auction. However, this could have provided an indication to TRAI for fixing the reserve prices for different bands in the future.

The liberalization of spectrum raises a further concern about the contiguity of spectrum bands. As discussed, spectrum was assigned to operators in tranches of 1–2 MHz based on an SBC, as and when it was available. Hence, the operators received spectrum in non-contiguous blocks. Figure 10.5 illustrates the assignment of 900 MHz and 1800 MHz spectrum. For 2G services that operate using GSM technology, the channel width is only 200 KHz and hence the non-contiguity is not a major constraint in providing services. The same case holds good for CDMA services which required a channel width of 1.25 MHz for CDMA 1X service. However, as the spectrum is liberalized, the operators can deploy any technology including IMT and IMT Advanced systems for providing 3G and

Table 10.18   The status of spectrum allocation post March 2013

S. No.	Description	1800 MHz	800 MHz	900 MHz
1	No. of quashed licences	98	24	0
2	Total quantity of spectrum vacated due to cancellation of licences	413.6 MHz	60 MHz	0 MHz
3	Quantum of spectrum put for auction in November 2012	295 MHz	95 MHz[2]	0 MHz
4	Quantum of spectrum sold in November 2012 auction	127.5 MHz	0 MHz	0 MHz
5	Quantum of spectrum put for auction in March 2013	57.50 MHz	95 MHz	42.6 MHz
6	Spectrum sold in March 2013 auction	Nil	30 MHz	Nil
7	Quantum of spectrum required to be put in to auction as per Supreme Court order {= (2) − {(4)+(6)}	286.10 MHz	40[3] MHz	0 MHz

Source: TRAI (2013).

[2] As per the auction design explained in the previous section, three blocks of 1.25 MHz were put up for action. In addition, an additional block of 1.25 MHz was made available as a top-up for the new entrants. In the Punjab and Andhra Pradesh, only two blocks of 1.25 MHz were put up due to non-availability. No block was put up in Rajasthan due to non-availability of spectrum. Due to the minimum block size requirements, the spectrum put up for auction is in fact more than the amount of spectrum assigned for quashed licensees.

[3] Due to assignment of more spectrum than that released due to cancellation of licences as per the 800 MHz auction guidelines, this is calculated only for LSAs in which the spectrum sold was less than that released due to cancellation. Hence there is discrepancy.

Table 10.19 Status of spectrum assignment across LSAs post March 2013 auction

Service area	Spectrum vacated due to cancellation of licences		Spectrum sold		Difference between spectrum vacated due to cancellation of licences and spectrum sold	
	1800 MHz	800 MHz	1800 MHz in Nov 2012	800 MHz in Mar 2013	1800 MHz	800 MHz
Delhi	4.4	2.5		3.75	4.4	
Kolkata	17.6	2.5	5	3.75	12.6	
Mumbai	13.2	2.5			13.2	2.5
Andhra Pradesh	22	2.5	5		17	2.5
Gujarat	17.6	2.5	10	3.75	7.6	
Karnataka	22	2.5		3.75	22	
Maharashtra	22	2.5	6.25		15.75	2.5
Tamil Nadu & Chennai	22	2.5	5	3.75	17	
Haryana	22	2.5	7.5		14.5	2.5
Kerala	17.6	2.5	1.25	3.75	16.35	
Madhya Pradesh	17.6	2.5	7.5		10.1	2.5

Punjab	17.6	2.5	1.25		2.5	16.35
Rajasthan	17.6	0			0	17.6
Uttra Pradesh (E)	17.6	2.5	11.25		2.5	6.35
Uttra Pradesh (W)	17.6	2.5	12.5	3.75	2.5	5.1
West Bengal, Andaman & Nicobar	17.6	2.5	8.75	3.75		8.85
Assam	22	5	8.75		5	13.25
Bihar	22	2.5	13.75		2.5	8.25
Himachal Pradesh	17.6	2.5	1.25		2.5	16.35
Jammu & Kashmir	22	5	7.5		5	14.5
North East	22	5	7.5		5	14.5
Odisha	22	2.5	7.5		2.5	14.5
Total	**413.6**	**60**	**127.5**	**30**	**40**	**286.1**

Source: TRAI (2013).

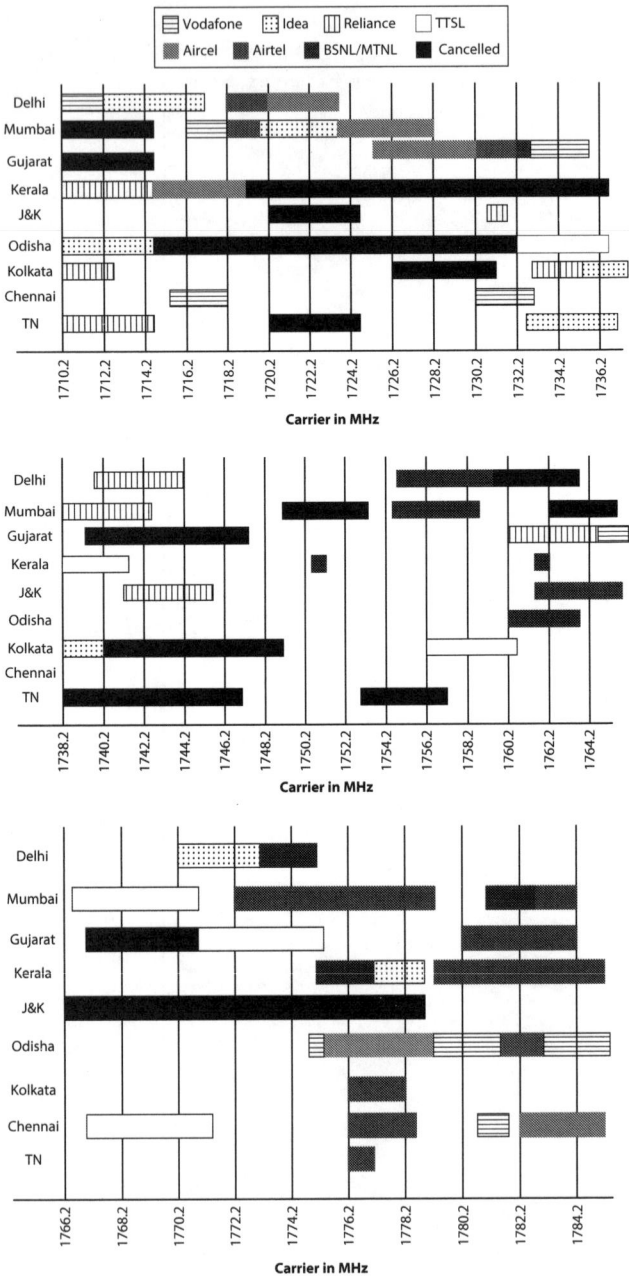

Figure 10.5 Carrier distribution of GSM spectrum allotted to telecom service providers licence area network wise

Source: TRAI (2012).

4G services. These require a minimum contiguous channel width of 5 MHz for efficient utilization. Hence, re-configuration of frequencies is required in the liberalized regime to provide contiguous spectrum blocks to the operators. This is all the more important in the refarming of the 900 MHz.

The 5G technologies such as LTE-Advanced use a technique known as carrier aggregation to aggregate both contiguous as well as non-contiguous channels either within or across bands to enhance throughput as explained in Chapter 1. Hence, the mobile operators need to move towards these technologies, especially in dense urban areas, to improve the spectral efficiency and to provide more data throughput.

Auction of 900 MHz and 1800 MHz in February 2014

The 900 MHz spectrum vacated at the close of the licence period in November 2014 in the three metros of Delhi, Mumbai, and Kolkata, and the 1800 MHz spectrum unsold after the 2013 auctions was put on the block in February 2014. Since there was no reservation of 1800 MHz block for refarming 900 MHz of the incumbent holdings, the amount of 1800 MHz put on the block increased. TRAI proposed about 37 per cent reduction in the base price for the 1800 MHz spectrum and about 60 per cent reduction for the 900 MHz. Significant aspects of this recommendations included uniform spectrum usage charges instead of the tiered rate that is being practiced and allowing trading of the spectrum that has been auctioned. However, the recommendations did not include a base price for 800 MHz as the roadmap for services that use technologies (for example, CDMA) in the 800 MHz is very unclear and the adoption of these technologies has been poor both by the operators and users. The base prices recommended by TRAI were increased by about 25 per cent by the Telecom Commission for both 900 MHz and 1800 MHz. Table 10.20 illustrates the details of the auction that started on 3 February 2014 and lasted for 10 days. Eight operators participated in the auction. The notable entrant was Reliance Jio which has pan-India spectrum in 2300 MHz for offering the broadband wireless access service. Since 2300 MHz is inefficient in terms of propagation characteristics and involves more base stations and

Table 10.20 Results of the 1800 MHz spectrum auction held in February 2014

| | 1800 MHz | | | | | |
LSA	Winning price (in Rs. crore) per block of 200 KHz	Winners (MHz)	Total spectrum assigned (pair of MHz)	Total bid amount (in Rs. crore)	Reserve price per block of 200 KHz (in Rs. crore)	Percentage increase from reserve price
Andhra Pradesh	32.60	Bharti (8.8); Voda (0.6); Idea (6); R Jio (5.8); Telenor (1.4)	22.60	3,684	32.60	0.00
Assam	7.22	R Jio (5.4); Telenor (6)	11.40	412	1.40	415.71
Bihar	8.62	Telenor (2.2)	2.20	95	7.40	16.49
Delhi	72.80	Bharti (7); Voda (8); Idea (0.6); R Jio (5.4)	21.00	7,644	43.80	66.21
Gujarat	47.56	Voda (4.4); Idea (1.6); R Jio (6)	12.00	2,854	28.60	66.29
Haryana	5.40	Voda (2.4); Idea (6)	8.40	227	5.40	0.00
Himachal Pradesh	1.20	Bharti (10.2)	10.20	61	1.20	0.00
Jammu & Kashmir	1.22	Bharti (2.6); Aircel (1.8)	4.40	27	1.00	22.00
Karnataka	31.00	Bharti (8.8); Voda (5); Idea (5); R Jio (5);	23.80	3,689	31.00	0.00
Kerala	10.40	Bharti (5); Voda (7); Idea (10); R Jio (5)	27.00	1,404	10.40	0.00

Kolkata	14.60	Bharti (5); Voda (8); R Jio (5)	18.00	1,314	14.60	0.00
Madhya Pradesh	10.08	Bharti (5.8); Idea (7); R Jio (6.4)	19.20	968	8.60	17.21
Maharashtra	58.07	Idea (9); R Jio (5)	14.00	4,065	34.60	67.83
Mumbai	54.40	Bharti (6); Voda (8.2); Idea (2); R Jio (6.6); RCOM (0.6)	23.40	6,365	41.40	31.40
North East	1.40	Bharti (7); Idea (5); R Jio (6.4); Aircel (1.8)	20.20	141	1.40	0.00
Odisha	3.20	Bharti (5); R Jio (5)	10.00	160	3.20	0.00
Punjab	10.80	Bharti(8.2); Voda(0.6); Idea (8)	16.80	907	10.80	0.00
Rajasthan	5.20	Bharti (8.2); Voda (0.8); Aircel (1.6)	10.60	276	5.20	0.00
Tamil Nadu	41.60	Bhart i(5); R Jio (6.2)	11.20	2,330	41.60	0.00
Uttar Pradesh (E)	12.80	Voda (4); Telenor (1.8); Aircel (1.8)	7.60	486	12.20	4.92
Uttar Pradesh (W)	18.99	Telenor (2)	2.00	190	12.40	53.15
West Bengal	4.92	Bharti (4.4); R Jio (5.6); Aircel (1.2)	11.20	276	4.20	17.14
Total			**307.20**	**37,573**		

Source: Authors' own.

hence increased capital and operating expenditure, it was expected that this operator would bid aggressively to acquire 900 MHz. Further it was expected that in the other circles, Reliance Jio would bid for 1800 MHz to provide GSM voice services.

The auction was successful compared to the damp squib of the previous two auctions with the government raking in Rs. 61,162 crore, about 25 per cent of the annual turnover of the telecom service industry. This is about 45 per cent more than what was guaranteed under the set reserve price. However, the winners are likely to exercise the deferred payment option under which only 33 per cent of the winning bid price for 1800 MHz and 25 per cent of the 900 MHz needs to be paid up front with a moratorium of 2 years and the remaining to be paid over 10 equal instalments.

Apart from the one-time fee, the operators will continue to pay annual spectrum usage charges. The EGoM approved a uniform spectrum charge of 5 per cent on the newly auctioned radio waves. The incumbents will pay a weighted charge including the tiered rate explained earlier for the administratively assigned spectrum. The BWA access providers will continue to pay 1 per cent of revenue as the spectrum usage charges for their holding of 2300 MHz spectrum as per the conditions laid down before the 2010 auction.

The following are the important takeaways from this auction:

(1) There was a tremendous pressure for the incumbent holders of 900 MHz to retain their spectrum holding to offer continued service to their subscribers. In both Delhi and Kolkata, Bharti Airtel and Vodafone, the incumbent holders, had to compete aggressively with Reliance Jio for acquiring 900 MHz. In Mumbai, one of the incumbent holders, Loop Telecom, did not participate in the auction. However, its holding of 8 MHz was also put up for auction. The effect is shown in the winning bid prices which on an average were 85 per cent more than the reserve price. In fact, the winning bid price for the metros was about 6 per cent more than the 3G winning bid prices witnessed in the 2010 auction.

(2) The acquisition of 1800 MHz varied across different circles, with much of the activity concentrated in Assam, Bihar, Mumbai, Maharashtra, UP(E), UP(W), and J&K with little activity in the other circles. In fact, the winning bid price was equal to the

Table 10.21 Results of the 900 MHz spectrum auction held in February 2014

	900 MHz					
LSA	Winning price (in Rs. crore) per block of 1 MHz	Winners (no. of blocks won)	Total no. of blocks assigned	Total spectrum assigned (pair of MHz)	Total bid amount (in Rs. crore)	Reserve price per block of 1 MHz (in Rs. crore)	Percentage increase from reserve price
Delhi	740.96	Bharti (6); Voda (5); Idea (5)	16	16	11,855	360	105.82
Kolkata	194.63	Bharti (7); Voda (7); Idea (5)	14	14	2,725	125	55.70
Mumbai	563.09	Bharti (5); Voda (11)	16	16	9,009	328	71.67
Total			**46**	**46**	**23,590**		

Source: Authors' own.

reserve price in 50 per cent of the circles. Reliance Jio which has 20 MHz of 2300 MHz spectrum pan-India won 1800 MHz blocks in 14 of the 22 circles. Apart from voice, this spectrum has an evolving ecosystem for the 4G technology, known as Frequency Division-Long Term Evolution (FD-LTE). However, the Time Division version of the LTE technology called TD-LTE is being deployed in the 2300 MHz band. Hence, it is expected that India will be one of the first countries to witness coexistence of FD and TD-LTE technologies and will witness dual mode handsets and dongles. The operators are likely to strengthen their 2300 MHz coverage by better propagating 1800 MHz for their 4G services.

(3) The incumbent holders are expected to migrate their existing voice subscribers from 900 MHz to 1800 MHz (possible as all handsets are multi-band ones) and provide 3G services (based on WCDMA/HSPA technology) in their reacquired 900 MHz. The 3G ecosystem in 900 MHz is well developed and mature enough for the operators to provide cost effective services. Moreover, these being metros, there is enough latent demand for high quality mobile broadband services; this demand is currently being inadequately served by the operators due to the limited 2100 MHz spectrum that they acquired in 2010. Interestingly, Bharti and Vodafone also acquired 1800 MHz in sufficient quantities for the earlier mentioned migration plan. Hence, we can expect 3G roll-out in a big way in Delhi, Mumbai, and Kolkata.

(4) It is one of the rare auctions in which a spectrum block as narrow as 200 MHz (paired) was auctioned in 1800 MHz. It is to be noted that 200 MHz is the minimum carrier requirement for GSM services. Though the auctioned spectrum is technology neutral, the block size is related to 2G technologies. For providing 3G/4G services, the operators need to have at least 5 MHz. Hence, the 1 MHz block auctioned in 900 MHz also is very low and the operator needed to bid for at least five blocks to provide 3G/4G services.

The results of this auction have important implications for future auctions. The constrained spectrum availability and lack of a clear roadmap for future spectrum allocation is one of the reasons for the high prices witnessed in 900 MHz. Moreover, the 900 MHz spectrum holding of the incumbents is once more coming up for renewal at the end of 2014. What should the action of the government and regulator be before the next auction?

(1) It is essential that the government provides a good roadmap for 2100 MHz and holds the auction either before or simultaneously with the 900 MHz auction so that the operators can bid more wisely. The 2100 MHz spectrum will continue to be the workhorse for broadband deployment for some time to come. The swapping deal long talked about between the Department of Defence and DoT on 1900 MHz from DoT to Defence versus 2100 MHz from Defence to DoT in each circle should be completed and enough 2100 MHz put on block to satisfy the 3G demands.

(2) Non-contiguous spectrum is the bane of operators. Figure 10.6 illustrates the non-contiguity of spectrum in the February 2014 auction in both 1800 MHz and 900 MHz. Non-contiguous spectrum tremendously decreases spectral efficiency, and is not suitable for providing broadband data services. The government and the regulator should immediately initiate actions to re-arrange discontinuous blocks of spectrum in the 1800 MHz band among the incumbent spectrum holders to create contiguous blocks, so that the entire acquired spectrum in the 1800 MHz band becomes suitable for providing 4G services. It also appears that defence is using some portion of the 1800 MHz band. Moving defence users to an alternate spectrum might help expedite the harmonization process and enable release of additional spectrum in the 1800 MHz band, which can also be put for assignment in the next auctions. The 2100 MHz spectrum along with the additional 1800 MHz spectrum can help prevent excessive price rise of the future 900 MHz band spectrum, thereby mitigating cost of services.

Though the notice inviting applications for the auction held in February 2014 indicated that spectrum trading and sharing would be allowed for the auctioned spectrum, detailed guidelines form the government are awaited even as of April 2014. Details of spectrum assignment across bands to operators in different LSAs are provided in Table 10.22.

Legend: Bharti [B] | Idea [I] | RJIO [R] | Telenor [T] | Vodafone [V] | Aircel [A] | RCOM [RC]

1800 MHz S.No Circle	Slot 1	Slot 2	Slot 3	Slot 4	Slot 5	Slot 6	Slot 7	Slot 8	Slot 9	Slot10
1 AP										
2 Assam										
3 Bihar										
4 Delhi										
5 Gujarat										
6 Haryana										
7 HP										
8 J&K										
9 Karnataka										
10 Kerala										
11 Kolkata										
12 MP										
13 Maha										
14 Mumbai										

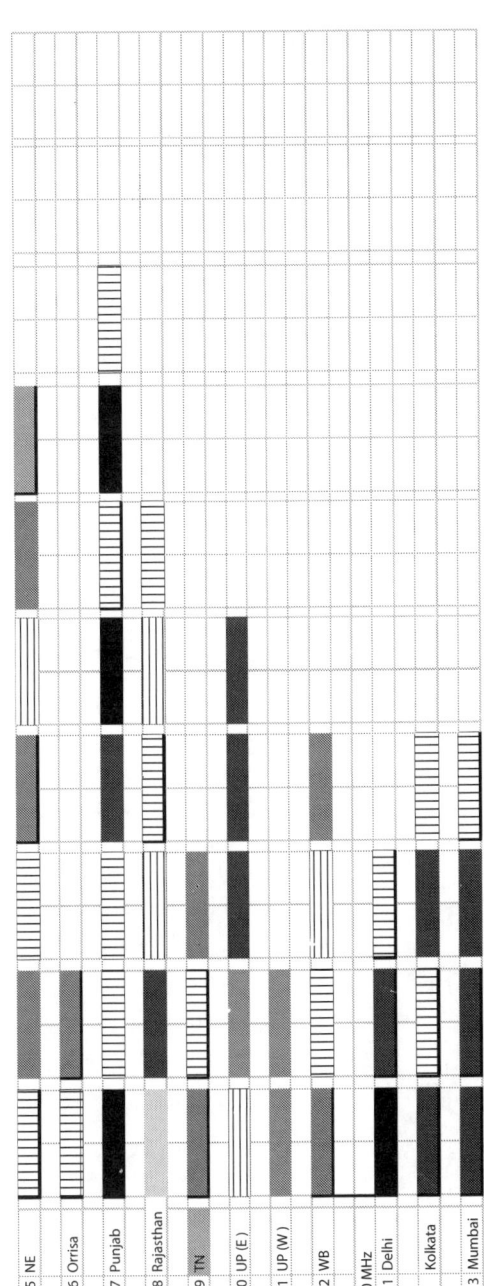

Figure 10.6 Non-contiguity of spectrum assignment across circles to various operators in the auction in February 2014

Source: Authors' own.

Table 10.22 Average spectrum assignment per operator in different LSAs as on July 2013

	800 MHz		900 MHz		1800 MHz	
	No. of spectrum holders	*Average amount (in MHz)*	*No. of spectrum holders*	*Average amount (in MHz)*	*No. of spectrum holders*	*Average amount (in MHz)*
Andhra Pradesh	3	4.58	3	6.73	9	6.09
Assam	2	2.50	4	4.65	9	4.33
Bihar	3	3.75	3	6.20	9	4.68
Delhi	4	4.38	3	5.33	6	7.10
Gujarat	4	3.75	3	6.73	10	4.46
Haryana	3	3.75	3	6.20	8	4.89
Himachal Pradesh	3	2.50	3	6.20	6	5.48
Jammu & Kashmir	2	2.50	3	6.20	6	4.18
Karnataka	4	3.75	3	6.73	9	5.67
Kerala	4	4.06	3	6.20	8	6.66
Kolkata	4	3.75	2	7.00	8	5.70
Madhya Pradesh	3	3.33	3	6.20	8	6.69
Maharashtra	3	3.75	3	6.73	9	5.27
Mumbai	3	4.17	2	8.00	9	6.71
North East	2	2.50	4	4.85	8	5.49
Odisha	3	2.92	3	6.20	9	4.03
Punjab	4	3.13	3	7.27	10	4.19
Rajasthan	4	3.75	3	6.20	8	4.78
Tamil Nadu	3	4.17	3	6.73	7	5.09
Uttar Pradesh (E)	3	3.75	3	6.20	9	5.01
Uttar Pradesh (W)	4	3.75	3	6.20	10	3.95
West Bengal	4	3.13	4	4.85	9	3.86
Pan India	**3**	**3.50**	**3**	**6.26**	**9**	**5.06**

Source: Authors' own.

[4] For unpaired frequency blocks assigned in 2300 MHz and 2600 MHz, half of the spectrum blocks assigned are taken for spectrum HHI calculations, since all the other spectrum holdings are paired.

2100 MHz		2300 MHz		Spectrum HHI[4]		
No. of spectrum holders	Average amount (in MHz)	No. of spectrum holders	Average amount (in MHz)	Total no. of spectrum holders	Paired	Including unpaired[5]
4	5.00	2	20.00	9	0.14	0.14
4	5.00	2	20.00	9	0.14	0.15
5	5.00	3	20.00	12	0.12	0.11
4	5.00	2	20.00	10	0.16	0.13
4	5.00	2	20.00	11	0.11	0.11
4	5.00	2	20.00	10	0.15	0.13
5	5.00	1	20.00	9	0.17	0.16
6	5.00	2	20.00	8	0.15	0.15
4	5.00	2	20.00	9	0.14	0.15
4	5.00	2	20.00	10	0.14	0.12
4	5.00	2	20.00	9	0.13	0.14
4	5.00	2	20.00	10	0.13	0.13
4	5.00	2	20.00	9	0.14	0.14
4	5.00	2	20.00		0.16	0.14
4	5.00	2	20.00	8	0.14	0.15
4	5.00	2	20.00	9	0.13	0.15
5	5.00	2	20.00	10	0.12	0.12
4	5.00	2	20.00	10	0.15	0.13
4	5.00	2	20.00	9	0.13	0.14
4	5.00	2	20.00	11	0.13	0.12
4	5.00	2	20.00	12	0.12	0.11
5	5.00	2	20.00	9	0.13	0.14
4	**5.00**	**2**	**20.00**	**10**	**0.14**	**0.13**

Spectrum HHI refers to the Herfindahl-Hirschman Index (HHI) of the spectrum holdings by various operators. It is computed as $HHI = \sum_{i=1}^{n} f_i^2$ where f_i refers to a fraction of spectrum holding by an operator in that ser-vice area and n refers to the total number of operators with spectrum holding

Following the announcements of uniform spectrum usage charges, new M&A guidelines were issued by the government in February 2014. The salient features of these guidelines are (DoT February 2014):

(1) The sector is liberalized by allowing 100 per cent FDI, more suitable for foreign companies such as Vodafone to invest in the sector.

(2) Trading of spectrum between spectrum holders has been allowed.

(3) The market share cap for the resultant entity has been increased from 35 to 50 per cent; however, market share has been defined both in terms of subscriber base and AGR. The conditions on the minimum number of operators in each LSA have been removed.

(4) The total spectrum holding cap of the resultant entity has been retained at 25 per cent as per the earlier norms; however, a new cap of 50 per cent for each band has been included.

(5) As per the UL guidelines, no licensee or its promoter(s) directly or indirectly shall have any beneficial interest in another licensee company holding 'Access Spectrum' in the same service area. This prevents cross equity holding by licensee or promoter(s) in a service area which was earlier capped at 10 per cent.

(6) The mergers of all licences/authorization in a service category are allowed that includes mergers of ISPs with access providers such as wireline or wireless operators.

The guideline is silent about spectrum sharing between operators and allowing MVNOs in the country. It is time that clarity emerges on spectrum sharing as these will provide the operators much needed flexibility.

Spectrum Pricing: An International Comparison

Table 10.23 gives the price of auctioned spectrum across different LSAs for different spectrum bands since 2001.

It is to be noted that the price for metros is multiple times more than category A circles in any of the auctions, indicating the focus of the bidders. Price in category A is about twice that in category B and that in category B is about thrice that in category C.

---

in that service area. Spectrum blocks in different frequencies are added up to compute the total spectrum in a LSA. HHI can range from 0 to 1; where 1 indicates that all spectrum is held by one operator and a value closer to zero indicates excessive fragmentation of spectrum across different operators.

Table 10.23 Comparison of spectrum prices across different LSAs for different bands

	Price (in Rs) per MHz per population					
	800 MHz	*900 MHz*	*1800 MHz*			*2100 MHz*
LSA	In 2013	In 2014	In 2001	In 2012	In 2014	In 2010
Andhra Aradesh			3.10	26.35	18.20	32.42
Assam			0.43	2.24	11.33	2.75
Bihar			0.21	2.86	3.22	3.22
Delhi	200.84	407.17	28.56		200.03	385.55
Gujarat	18.96		4.94	29.51	37.94	36.31
Haryana			2.34	14.72	10.39	18.11
Himachal Pradesh			0.42	8.61	8.07	10.59
Jammu & Kashmir			0.45	3.92	4.60	4.83
Karnataka			8.97		24.13	52.03
Kerala	9.25		2.91	14.43	13.96	17.75
Kolkata	36.17	117.38	13.42	56.43	44.03	69.43
Maharashtra			5.31	4.45	5.05	5.48
Madhya Pradesh			0.49	29.22	39.24	35.95
Mumbai		261.50	28.17		126.32	318.96
North East			0.42	5.12	4.93	6.30
Odisha			0.31	3.85	3.70	4.74
Punjab			13.69	18.60	18.15	22.89
Rajasthan			1.31		3.74	9.78
Tamil Nadu	22.00		7.63	34.32	28.35	42.23
Uttar Pradesh (E)			1.25	6.13	6.26	7.54
Uttar Pradesh (W)	5.04		0.77	7.86	8.44	9.67
West Bengal, Andaman & Nicobar			0.03	9.02	2.51	2.67
Metros	**118.50**	**262.02**	**23.38**	**56.43**	**123.46**	**257.98**
Category A	**20.48**		**5.99**	**23.66**	**22.73**	**33.70**
Category B	**7.14**		**2.85**	**14.28**	**12.84**	**15.55**
Category C			**0.37**	**4.43**	**5.97**	**5.41**

Source: Authors' own.

Table 10.24 compares the spectrum prices held in recent times across different countries with that in India's recent auction. The following can be observed from Table 10.24:

(1) The spectrum price in metros is much higher than international averages clearly indicating the skewed growth of mobile services and spectrum paucity in the country. Except for the auctions that were held in Europe for the UMTS band during 2001, spectrum prices across bands across different countries are about the price discovered for category B circles in India.

(2) While in most of the countries the lower frequency spectrum is much valued as compared to the higher frequency spectrum,

Table 10.24 Comparison of spectrum prices across different countries and that of India

Countries	Price (in Euros) per MHz per population		
	900 MHz in 2014	*1800 MHz in 2014*	*2100 MHz 2010*
India			
Metros	3.09	1.46	4.51
Category A		0.27	0.59
Category B		0.15	0.27
Category C		0.07	0.09
France	1.35 (800 MHz)		0.74
Germany	1.54 (800 MHz)		0.05 (2600 MHz); 0.23 (2100 MHz); 10.68 (2100 MHz in 2001)
Sweden	0.68 (800 MHz)		0.2 (2600 MHz)
UK			7.63 (2100 MHz in 2001)
US	1.32 (700 MHz in 2008)		0.22 (2100 MHz in 2006)
Benchmark prices (Ofcom 2012)	0.28–0.86	0.17–0.27	0.1-0.15

Source: Mölleryd and Markendahl (2012).

strangely in India the 2100 MHz band was valued more than 900 and 1800 MHz clearly indicating severe spectrum scarcity in 2010 for 3G services.

(3) Table 10.24 also presents the possibility of a 'winner's curse' in the current auction and hence the need for the government to get its act together for releasing enough spectrum by the end of 2014 when the 900 MHz spectrum is nearing the end of its life in all the other 19 circles.

The Possibilities of Cognitive Radio

The National Telecom Policy of 2012 (NTP 2012) envisions a liberalized spectrum regime in which spectrum is technology and service agnostic. Further, it indicates that there will be permission to pool, share or trade spectrum among the operators to enable optimal utilization of scarce spectrum resources. It also mentions the possible release of the 700 MHz digital dividend spectrum and the possible use of cognitive radio (CR) to utilize white spaces in different frequency bands. The technologies and standards that are being developed for CR as explained in earlier chapters, provide an opportunity for India to lead in their use in a manner that is unique to us and to solve spectrum problems that are also unique to us. Towards this we ask: 'Can we formulate, for once, ex-ante policies and guidelines for the early implementation of these clauses in NTP 2012 to exploit and nurture the associated technology developments?'

The Indian mobile communication service market is unique with a large number of operators, each operating in different frequency bands (800 MHz, 900 MHz, 1800 MHz, 2100 MHz, 2300 MHz) using different technologies (GSM, CDMA, WCDMA, TD-LTE). As mentioned earlier, 700 MHz is expected to join the list and FD-LTE is expected to be deployed in this band. This forms a true test bed for CR. Further, allowing for technology agnostic use of spectrum as indicated in NTP 2012, allows spectrum owners to dynamically lease between themselves for short durations. It is time to go beyond the intra-circle and inter-circle roaming/sharing that we witness today, to a full-fledged cross-band, cross-operator, cross-circle sharing of spectrum—statically through contracts, and dynamically using CR-type technologies. We strongly believe that this will overcome the coverage and capacity constraints that we see today. A typical use case could involve an irritated subscriber whose native operator's network signal strength deteriorates rapidly in a specific location. He can be switched to another operator whose network signal strength is good by leasing spectrum on a short-term basis from the second operator. This will avoid the dreadful call drops.

The increasing availability of GSM/CDMA dual SIM handsets indicates that this scenario is well within the realm of technical possibility. The handset and network capabilities for cross-band sharing are already present and limited only by the associated economics. This also gives opportunities for specialized spectrum brokers to set up trading exchanges for spectrum where the operators can statically or dynamically trade spectrum, in the same manner as wholesale Internet/international bandwidth trading. It may be noted that such bandwidth exchanges for national/international long-distance services have been around for a while.

Though CR and CR-inspired technologies are in an infant stage of deployment in the world, India is best poised to leverage the developments in this field to create new technology scenarios and hence new opportunities. Given the size of the market, the competition level, and the supply-side constraints a proactive approach starting now can put us in a leadership position (Sridhar and Manjunath 2012).

An aggressive proactive approach will have implications for both the operators and regulators. For the secondary market in mobile spectrum to legally operate, guidelines for implementing the NTP 2012 policy clauses need to be released soon. Indian manufacturers and software companies should also view this as an opportunity to develop products with the capabilities mentioned earlier and be a pioneer in this space given the economies of scale that the Indian market provides. This is also in tune with the objective of NTP 2012 of promoting innovation and indigenous R&D and manufacturing to serve the domestic market.

Informal discussions indicate that the operators have the means and inclination to take advantage of a secondary market in mobile spectrum. Manufacturers can provide affordable handsets for dynamic spectrum access. The command and control approach of spectrum management practiced by our government and the associated ex-post policies have only taken us thus far. To reap the full benefits of wireless broadband what we need now is proactive policies and aggressive competition.

Spectrum for Research and Development

India has been the hub of global outsourcing post the Y2K era and has steadily made strides into the niche space of engineering R&D services (ERD) in the past decade. Telecom, consumer electronics, automotive, and aerospace contribute the most to this segment with revenues touching $13 billion, and expected to reach about $45 billion by 2020. Compared to the IT services industry which bets on cost arbitrage, ERD thrives on innovation, deep domain knowledge, intensive technical skills, and considerable business risk exposure. In the late 1990s and early 2000s, a number of niche Indian ERD firms along with large IT services companies sensed opportunities and captured this outsourcing market. Figure 10.7 illustrates the ERD market space (NASSCOM 2010).

As can be seen from Figure 10.7, telecom contributes to a significant portion of the ERD outsourcing market. Apart from the third party independent service providers, most of the multinationals have

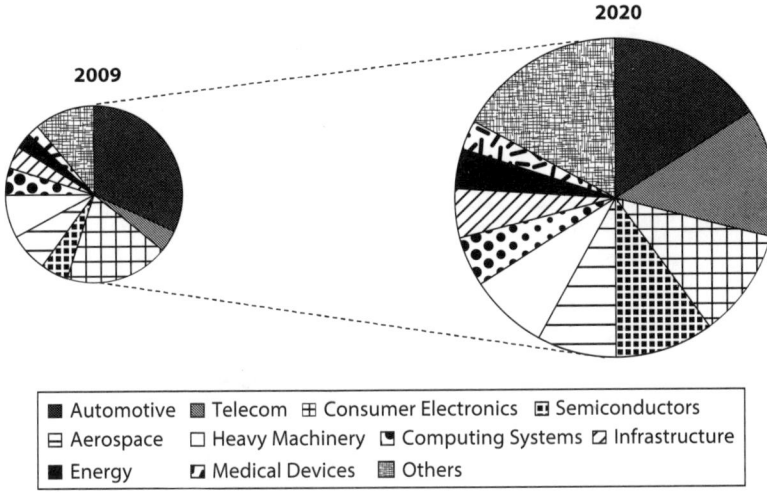

■ Automotive ▨ Telecom ⊞ Consumer Electronics ▧ Semiconductors
⊟ Aerospace ☐ Heavy Machinery ◪ Computing Systems ▨ Infrastructure
■ Energy ◪ Medical Devices ▦ Others

Figure 10.7 Indian ER&D offshoring market
Source: Adapted from NASSCOM (2010).

set up their captive ERD centres in India (for example, in Bangalore, Gurgaon, Mumbai, Hyderabad, and Pune) for product development and testing. These activities require the allocation of spectrum to be used in the respective labs of ERD firms. Figure 10.8 illustrates the different types of licences issued for experimental use for development and testing.

The licences given for experimental use for development and testing are broadly classified into two types: (a) radiating, and (b) non-radiating. In the non-radiating type, all the RF equipment needs to be connected using RF cables and connectors so that no radiation is emitted while in use. The Regional Licensing Offices (RLO) of the WPC issue these licences. A point frequency is assigned by WPC and the experimental set up is expected to use these point frequencies in a non-radiating manner. However, in case of a radiating licence, the procedure is more complex. First, only frequencies cleared under NFAP are allocated. The labs are expected to have RF shielding and the equipment tuned to low power so that no radiation is emitted outside.

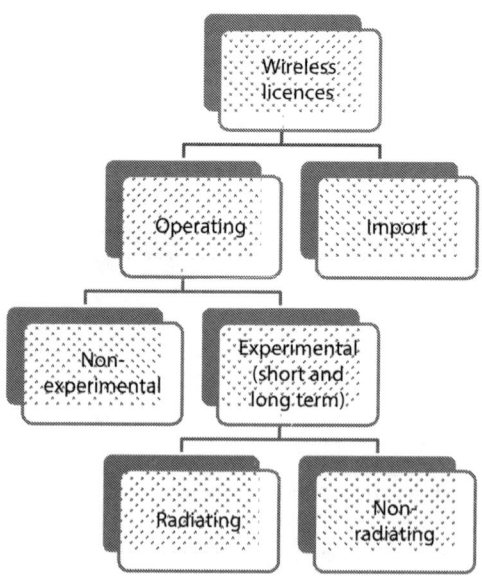

Figure 10.8   Types of licence issued for ERD and testing
Source: Authors' own.

Following are the challenges normally faced by the ERD firms:

(1) Most of the ERD firms work on future products, frequencies of which are not cleared under the NFAP. Examples include Terrestrial Trunked Radio (TETRA) in 400 MHz–430 MHz, 2.7 GHz, and 3.3 GHz for wireless broadband 4G services. Since the government holds the spectrum and does not clear them for usage, the ERD firms face difficulty in development and testing.

(2) Even for the available frequencies, the ERD firms need to be very cautious about radiation spilling outside their labs and causing harmful interference to existing commercial services.

(3) The experimental spectrum is often for a short term (6 months lasting up to about 2 years) and is administratively priced. However, in the event of the Supreme Court decision, there is a debate as to whether spectrum for experimental use needs to be market price based. If the price of spectrum increases, it will undermine India and Indian ERD firms as a low cost off-shore destination for advanced ERD product development and testing.

Spectrum for Development

Mobile access is likely to be the entry point of many less privileged citizens to the Internet. It is incumbent upon the government to fashion spectrum policy as well as complementary policies on content and licensing to facilitate the mainstreaming of Internet access for the population at large. As mentioned in Chapter 2, allocating large blocks of prime spectrum in rural areas for common use, and smaller blocks in urban areas will allow the growth of local ISPs that can bring about the proliferation of the Internet. Such rural ISPs should be charged a low licence fee in order to keep barriers to entry low. Success will depend on inculcating the necessary skills in the rural population and developing content and applications tailored to local use. The diffusion of the Internet is a far greater challenge than the spread of voice telephony as it involves acceptance of a new technology, and a certain level of basic literacy and relevant content. Encouraging local entrepreneurship is the way forward to achieve the multi-pronged challenges.

Conclusion

The Indian story is unique as it is strewn with legacy issues of the command and control era of spectrum management. The markets are highly fragmented and mired in litigation. The government navigates through the complex issues of technology and policy within the legacy of the past with varying degrees of adeptness and foresight. The good news is that the Indian mobile market is vibrant with intense competition and a large scalable subscriber base. Hence, India is the true testing ground, much like the US, for experimenting with newer ways of managing spectrum. It requires the will of all stakeholders concerned, the government as the policymaker, the regulator TRAI, and the operators to usher in a new era of spectrum management. As a summary, Table 10.25 indicates the tumultuous course of sequences in the Indian mobile sector.

Table 10.25 Important milestones in the Indian mobile sector

Year	Event
1991	DoT began the process of introducing private participation in the sector by inviting bids for two licences in each of the metros (Delhi, Mumbai, Kolkata, and Chennai)
31 July 1995	First cellular mobile service was started in Kolkata by Modi Telstra's Mobile Netservice. Bids were invited and two licences in the remaining 18 LSAs of the country were awarded using a single-stage sealed bid auction
1999	DoT introduced the migration to revenue-share based licence fee
2001	Government operators (BSNL and MTNL) were given the third mobile operator licence. The government issued guidelines for the fourth cellular licence. One licence in each LSA was awarded using a multi-stage auction.
2003	TRAI recommended fixed line service providers be allowed to provide Wireless Local Loop with Limited Mobility (WLL-M) service. On 31 October 2003, a council of ministers approved the UASL and on 11 November 2003, UASL guidelines were introduced

2005	TRAI made a comprehensive recommendation on spectrum allocation and pricing on 13 May 2005. On 14 December 2005, DoT modified the UASL guidelines removing the restrictions on the number of licensees in a service area
2006	On 23 February 2006, the Prime Minister approved the constitution of a group of ministers, consisting of the ministers of defence, home affairs, finance, parliamentary affairs, information and broadcasting, and C&IT, to look into issues relating to vacation of spectrum
2007	Government awarded new licences on a fixed fee basis to a number of players. Additionally under the 'cross over spectrum' policy, CDMA and GSM operators were made eligible to use both the technologies under the same licence DoT fixed 1 October 2007 as the last date for receiving applications for UASL; then through press notes, it was pre-poned to 25 September 2007
2008	Applications received between March 2006 and 25 September 2007 were issued letters of intent (LoIs) simultaneously on 10 January 2008; 122 UAS licences were issued using a flawed first-come-first-serve basis
2009	Government owned BSNL and MTNL were issued spectrum in 2600 MHz band for BWA services
Apr–Jun, 2010	3G and broadband wireless access spectrum auction: 3G auction conducted during 9 April 2010 and 19 May 2010; the BWA auction took place during 24 May 2010 and 11 June 2010.
2012	On 2 February 2012, the Supreme Court of India declared, 'the licences granted to the private respondents on or after 10 January 2008 and subsequent allocation of spectrum to the licensees are declared illegal and are quashed'. As per DoT reference TRAI made its recommendations on auction of spectrum subsequent to the court order on 23 April 2012 and a further revision on 12 May 2012 (TRAI 2012) Auction for 1800 MHz and 800 MHz was conducted during 12 November 2012 and 14 November 2012

(Continued)

Table 10.25 *(Continued)*

Year	Event
2013	On 7 January 2013 the EGoM decided on the guidelines for the second round of auction for 1800 MHz, 800 MHz as well as 900 MHz
	On 15 February 2013, the Supreme Court ordered that the entire spectrum vacated due to quashing of licences be auctioned without further delay
	On 11 March 2013, the second round of auction was completed
	On 23 July 2013, TRAI released its consultation paper on valuation and reserve price for spectrum subsequent to DoT reference (TRAI 2013)
	In August 2013, the unified licence guidelines were issued
2014	The EGoM announced uniform spectrum usage charges on 27 January 2014
	Auction of spectrum for 900 MHz and 1800 MHz started on 3 February 2014 and concluded on 13 February 2014
	New guidelines for M&As announced on 20 February 2014

Source: Authors' own.

References

Airtel. (2010). *Annual Financial Statements*. Available at: www.airtel.in. Accessed on 12 May 2013.

Alden, J. (2012). 'Exploring the Value and Economic Valuation of Spectrum.' Report prepared for the ITU. Available at: http://www.itu.int/ITU-D/treg/broadband/ITU-BB-Reports_SpectrumValue.pdf. Accessed on 31 October 2012.

Desai, A. (2006). *India's Telecommunications Industry: History, Analysis, Diagnosis*. New Delhi: Sage Publications.

Department of Telecommunications (DoT). (May, 2009). *Report of Committee for Allocation of Access (GSM/CDMA) Spectrum and Pricing*, Department of Telecommunications, Ministry of Communications and IT, Government of India, June 2008–May 2009 (referred to as Subodh Kumar Committee).

Department of Telecommunications (DoT). (23 Oct 2009). *Auction of 3G and BWA Spectrum: Revised Information Memorandum.* Available at: www.dot.gov.in. Accessed on 10 November 2009.

———. (2012). 'Auction of Spectrum in 1800 MHz and 800 MHz band: The Information Memorandum.' Available at: http://www.dot.gov.in.

———. (2013). 'License agreement for unified license.' Available at: http://www.dot.gov.in. Accessed on 10 November 2013.

Investment Information and Credit Rating Agency (ICRA). (2002). *Industrial Watch Series: The Indian Telecommunication Industry.* Mumbai: ICRA.

Klemperer, P. (2002). What really matters in auction design. *Journal of Economic Perspectives* 16(1): 169–89.

Mölleryd, B.G., and J. Markendahl. (2012). 'Valuation of spectrum for mobile broadband services- The case of Sweden and India.' Paper submitted to the regional ITS India Conference 2012 (22–24 February, New Delhi). Available at: http://www.its2012india.com/topics/Spectrum%20and%20Technology/ValuationofSpectrumfor MobileBroadbandServicesTheCaseofSwedenandIndia.pdf. Accessed on 9 November 2012.

National Association of Software and Services Companies (NASSCOM). (2010). *Global ER&D: Accelerating Engineering with Indian Innovation.* New Delhi.

Ofcom. (2012). 'Spectrum Value of 800 MHz, 1800 MHz and 2.6GHz: A DotEcon and Aetha Report for Ofcom.' Available at: http://stakeholders.ofcom.org.uk/binaries/consultations/award-800mhz/statement/spectrum-value.pdf. Accessed on 17 February 2013.

Singhvi, G.S., and Ganguly, A.K. (2 February 2012). Judgement: Supreme Court of India. Available at: http://supremecourtofindia.nic.in/outtoday/39041.pdf. Accessed on 10 August 2013.

Sridhar, V. (2006a). 'A battle of widths', *Financial Express*, 6 July.

———. (2012). '*The Telecom Revolution: Technology, Regulation and Policy*', New Delhi: Oxford University Press India.

———. (2013). 'Maximizing spectrum revenue', *Business Standard*, 5 March.

Sridhar, Kala, and V. Sridhar. (2010). 'One paisa revolution', *Financial Express*, 4 August.

Sridhar, V., and D. Manjunath. (2012). 'For a dynamic spectrum regime', *Economic Times*, 17 July.

Swamy, S. (2011). *2G Spectrum scam*. New Delhi: Har-Anand Publications Pvt. Ltd.

Telecommunications Regulatory Authority of India (TRAI). (2005). 'Recommendations on spectrum related issues. Retrieved from Telecom Regulatory Authority of India.' Available at: http://www.trai.gov.in/.

————. (2006). 'TRAI issues amendment to interconnect usage charges regulation.' Available at: http://www.trai.gov.in. Accessed on 29 April 2006.

————. (2007). 'Draft recommendations on growth of broadband.' Available at: http://www.trai.gov.in. Accessed on 20 September 2007.

————. (2012a). 'Recommendations on auction of spectrum.' Available at: http://www.trai.gov.in. Accessed on 15 December 2012.

————. (2013a). 'Consultation paper on valuation and reserve price of spectrum.' Available at: http://www.trai.gov.in. Accessed on 7 December 2013.

Wireless Planning Committee (WPC). (2008). 'Subscriber base criterion for allotment of GSM spectrum. Retrieved from Wireless Planning and Coordination Wing.' Available at: http://www.wpc.dot.gov.inS. Accessed on 8 December 2013.

11

Spectrum for Broadcasting Services

I love to think of nature as an unlimited broadcasting station, through which God speaks to us every hour, if we will only tune in.
—George Washington Carver (1864–1943)

Introduction

Radio and television broadcasting and mobile services have one thing in common—the use of spectrum for transfer of information. However, there are significant differences in the manner in which they use spectrum for information transfer. Radio and TV are traditionally one-way broadcasts with one transmitter transmitting over a large area, often the entire city while mobile services are two-way communications with transceivers in each micro-cell site. Managing interference in radio and TV transmissions is less complex because of its one-way broadcast nature, as opposed to mobile services where two-way transmission between handsets and base stations often involves tricky interference management. Cell re-use is non-existent in terrestrial radio and TV broadcasting as they operate at lower frequencies which have better propagation characteristics to cover entire cities with just one transmitter. However, both these services require allocation of spectrum and its effective management. In relation to

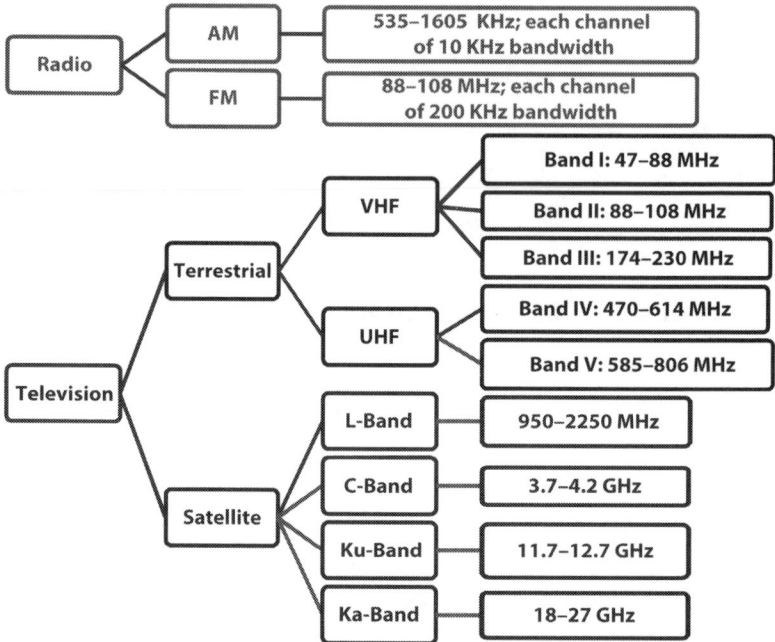

Figure 11.1   Spectrum used in radio and television broadcasting
Source: Authors' own.

the pricing of spectrum for broadcasting and mobile services, it is relevant that the use of community radio and TV has often been thought of as a public good and hence subjected to lower taxation compared to commercial mobile services which are even treated as a 'premium' service in some developing countries.

In this chapter we explore the radio and TV broadcasting services; their use of radio spectrum and the associated regulatory and policy dimensions. Figure 11.1 illustrates the various spectrum bands used in radio and TV broadcasting.

Radio Broadcasting

The first and foremost use of JC Bose and Marconi's discovery of using radio spectrum to transmit information over a long distance was in broadcast radio in the early part of the twentieth century.

During this time, radio spectrum was in abundance as compared to demand and 'radio hackers' who operated amateur radios (ham radios) started broadcasting over the ether as explained in Chapter 1. The era of spectrum non-regulation in the US ended in the wake of the *Titanic* disaster in 1912 which was ascribed to the 'chaos of spectrum'. The result was the Radio Act of 1912 which authorized the secretary of commerce to licence users of equipment that communicated via spectrum.

Formal radio broadcasting began in the US in November 1920 and developed rapidly (Hazlett 1990). By the end of 1922, there existed 576 broadcast stations in the US. Each had received a federal licence (at zero charge) from the secretary of commerce. An excess demand for zero-priced broadcast spectrum developed causing the classic 'tragedy of the commons'. Secretary Herbert Hoover (an engineer by training) pointedly withheld additional licences on the grounds that interference would otherwise result. However, when courts balked at Hoover's assertion of authority on the issue of broadcasting licences, Congress finally sought to establish a system of excludable property rights through the Radio Act of 1927 which established the Federal Radio Commission and gave it a broad jurisdiction to regulate access to spectrum under the broad principles of a general 'public interest' standard. When the Communications Act of 1934 was passed, the regulatory functions were transferred to the newly established Federal Communications Commission (FCC).

In emerging economies, radio is still the most popular and affordable means for mass communication, entertainment, and education. The terminal devices are affordable and portable. Even low-end mobile phones have FM tuners. Though the radio dipped in popularity after the diffusion of television, it has regained much lost ground due to the government's initiatives in allowing private parties to enter this segment.

In India, radio coverage is available in Amplitude Modulation (AM) (both short wave and medium wave), and Frequency Modulation (FM) modes.[1] In terms of coverage, AM broadcast covers almost 99 per cent of the Indian population and 100 per cent of

---

[1] AM works by varying the strength of the transmitted signal in relation to the information being sent. In contrast, the frequency is varied in FM.

the area, and FM covers about 40 per cent of the population and 25 per cent of the geographical area of the country (TRAI 2008).

As an initial step towards consolidating efforts in public service broadcasting, the Government of India created Prasar Bharati, a statutory autonomous body established under the Prasar Bharati Act. Prasar Bharati's board came into existence in November 1997. Prasar Bharati is a public-service broadcaster in the country. The objective of public service broadcasting is to be achieved through All India Radio (for public radio) and Doordarshan (for public television) which were earlier working as independent media units under the Ministry of Information and Broadcasting.

It is to be noted that in case of FM radio, the geographical jurisdiction of operators is relatively small compared to the much wider telecom service areas in case of mobile services. One reason for city-based licensing and allocation of spectrum is the localized nature of radio content compared to the ubiquitous mobile services. Another important factor is that radio consumers do not pay for listening, unlike mobile subscribers where the caller pays and TV viewers who often pay subscription charges. Advertisements are the only monetization model for radio services. Hence, the industry is extremely cost conscious. This makes determining spectrum price and the annual licence fee critical in the success of the business model.

The cities in which licence and spectrum are assigned are categorized as A+, A, B, C, and D based on the population (see Table 11.1). This practice is similar to the service area categorization in the case of mobile services.

Table 11.1 Categorization of cities for FM broadcasting service

Category of cities	Population criteria for classification of cities
A+	Metros
A	Population >20 lakhs
B	Population >10 lakhs and up to 20 lakhs
C	Population >3 lakhs and up to 10 lakhs
D	Population >1 lakhs and up to 3 lakhs

Source: TRAI (2008).

Table 11.2 Radio frequency allocation for FM broadcasting in NFAP 2011

Clause	Frequency
IND 13	FM broadcasting: 87–91.5 MHz and 95–100 MHz on a case-by-case basis
IND 14	FM broadcasting: 91.5–95 MHz
IND 15	Spot frequencies for private FM broadcasting: 88–100 MHz and 103.8–108 MHz

Source: WPC (2011).

The spectrum blocks allotted for FM as per the National Frequency Allocation Plan 2011 are given in Table 11.2.

However, as in mobile services, one of the issues, especially in A+ and A cities is the scarcity of FM spectrum in relation to the demand. FM radio operators have been requesting the government to reduce the inter-channel spacing from the current 80 KHz. TRAI in its recommendations released in April 2012 recommended reducing the inter-channel spacing to 40 KHz (TRAI 2012a) so that more spectrum can be given to the operators.

The licence and spectrum for FM broadcasting was carried out in three phases as outlined now.

Phase 1: 1999–2000

The government's policy objective for radio in the Ninth Five Year Plan was improving its variety of content and technical quality. On the technology front, the focus shifted from Amplitude Modulated (AM) transmissions to Frequency Modulated (FM) transmission as the latter has a much better performance in the presence of noise. In line with the policy of liberalization and reforms followed by the government since 1991, during the Ninth Five Year Plan it allowed Indian companies to set up private FM radio stations. Until then the government entity All India Radio was the sole radio broadcaster in the country. In May 2000, the government started the first phase and identified 108 frequencies in the FM spectrum (87–108 MHz)

Table 11.3 Details of the FM auction

	First phase	Second phase	Third phase policy guidelines
Method of allocation	Single-stage auction	Two-stage auction; first stage for eligibility and second stage for financial bidding; permission shall be granted on the basis of a one-time entry fee (OTEF) quoted by the bidders (closed tender system)	Simultaneous ascending auction as in case of 3G and BWA
Number of channels and cities	108 channels and 40 cities auctioned; finally only 21 channels in 12 cities became operational	337 channels encompassing 91 cities; finally 245 channels spanning over 87 cities were licenced; all cities with population >3 lakhs	Category A+: 9 to 11 channels; Category A: 6; Category B: 4; Category C: 4; Category D: 3
Reserve price and licence fee	Reserve Price: A+ city: Rs. 125 lakh; A: Rs. 100 lakh; B: Rs. 75 lakh; C: Rs. 50 lakh; D: Rs. 20 lakh	Annual licence fee: 4% of gross revenue for each year or 10% of the reserve one-time entry fee quoted during the auction	Same as in phase II

Source: Sridhar (2012).

for auction across 40 cities in the country. Details of the auction are given in Table 11.3.

Though bids were received for 101 frequencies, only 21 channels spanning 12 cities became operational. One of the main reasons cited was the high licence fee resulting in unviable business models. To analyse the outcome and define the path forward, the Radio Broadcasting Policy Committee was set up in July 2003.

Phase II: 2004–5

In February 2004, TRAI was asked by the Ministry of Information and Broadcasting to give guidelines for Phase II of private FM radio licensing. In August 2004, TRAI submitted its recommendations. Based on these recommendations and the report of the Radio Policy Committee, the Phase II policy was announced in July 2005, placing for bid 337 channels encompassing 91 cities. After scrutiny, 245 channels spanning 87 cities were given licences for FM broadcasting. The licences were valid for 10 years.

In order to reduce proliferation of towers, it was made mandatory for all Phase II operators to co-locate transmission facilities in all the cities except where new towers were to be constructed by the ministry. Every applicant was allowed to bid for only one channel per city with the further stipulation that the total number of channels allocated to an applicant and its related entities would not exceed the overall limit of 15 per cent of the total channels allocated in India. This clause was put in to reduce monopolization of content.

At the end of the two phases, spectrum for 92 Indian cities was picked up by private FM broadcasters, all with populations more than 3 lakhs. With the objective of covering cities with a population below 3 lakhs where there is demand for radio services, TRAI released a consultation paper in January 2008 for the third phase of FM broadcasting and made its recommendations. A majority of the recommendations were accepted by the government. A summary of the recommendations on Phase III is given in Table 11.3. EGoM in its meeting held on 6 March 2013 decided to charge migration fee from the existing operators on their migration from Phase-II to Phase-III and the process of migration is expected to be complete by June 2014.

Phase III: 2013 onwards

The grounds for further expansion of private FM radio broadcasting by bringing in the Phase III policy are:

(1) The FM Phase II policy has been well accepted and has resulted in huge growth in the FM radio industry.
(2) A huge unmet demand exists for FM radio in many cities which still remain uncovered by private FM radio broadcasting. Only state capitals and a limited number of cities with a population of 3 lakhs and above were taken up for bidding during the first two phases of FM radio broadcasting.
(3) Border areas, particularly in J&K, the north eastern states, and island territories, are largely missing from the FM map. Even those places that were put up for auction could not find takers due to poor viability. There is a need for promoting private FM radio in border areas to draw people to listen to Indian radio channels and to check cross-border propaganda. Similar initiatives are required for island territories. There are 97 vacant channels from Phase II which could not be auctioned due to various reasons.
(4) There is scope for further utilization of the frequency spectrum earmarked for FM broadcasting along with the generation of additional revenue for the government.

The FM Phase III policy extends FM radio services to about 227 new cities, in addition to currently covered ones. This Phase III policy will result in coverage of all cities with a population of one lakh and above with private FM radio channels. Table 11.4 gives the number of channels to be allocated in each type of city.

The Phase III design of the auction and the approach of fixing the reserve price are very similar to the 1800 MHz spectrum auction held in November 2012 for mobile services. While the government expects to garner about Rs. 1,500 crore in the auction, the high reserve prices may prove to be a dampener.

The main difference between spectrum policy for radio and commercial mobile services is that the radio spectrum policy is not moving towards the liberalization of spectrum. The reason is that the technology has not evolved to make the FM spectrum amenable for

Table 11.4   Allocation of FM radio channels in each city category in Phase III allocation in India

City category	No. of channels
A+	9 to 11
A	6
B	4
C	4
D	3

Source: TRAI (2008).

services other than radio communication. However, some exciting new developments in low-power broadcasting could make flexible spectrum use in radio a reality in the near future.

Television Broadcasting

There are two types of television broadcasting. Terrestrial TV or over-the-air TV, the original mode of broadcast uses low frequency spectrum to transmit directly to a home or, more recently, to a local cable operator who sends the signal through co-axial or fibre optic cables to homes. Satellite TV uses high-frequency spectrum to broadcast to satellites which beam the signal either to a local cable operator or directly to homes. Terrestrial broadcast traditionally used analogue signals, while satellite TV which emerged later, has used digital signals for most of its life. The migration of terrestrial broadcasts to digital signals is freeing up large amounts of spectrum which can be used for commercial mobile services. This is the 'digital dividend' spectrum. Based on the type of spectrum used for broadcast and the presence or absence of cable in the transmission to the home, we can place the different types of TV transmissions into four types (Table 11.5). Irrespective of the mode in which signals are broadcast, a cable operator can transmit digital or analogue signals to homes. Of late, cable operators have also begun the process of moving from analogue to digital transmissions, requiring the receiver in a home to have a digital tuner. The cable TV value chain consists of a multi-system operator (MSO) who aggregates signals from the broadcasters and

Table 11.5 Models of TV broadcasts

Broadcasting modes	Terrestrial (low frequency in 700 MHz)		Satellite (high frequency in L, C, Ku, and Ka bands)	
Type of transmission	Analogue	Digital	Analogue	Digital
Direct wireless access	Rooftop antennas with analogue TV (traditional over-the-air broadcasting)	Rooftop antennas with digital TV	Rooftop antennas with analogue TV and tuner (not allowed in certain countries due to large antenna size)	Direct-to-Home (DTH)
Wireline access through cable TV	Signals sent through cables and viewed with analogue TV	Signals sent through cables and viewed with digital set top box or digital TV	Signals sent through cables and viewed with analogue TV	Signals sent through cables and viewed with a digital set top box or digital TV

Source: Authors' own.

sends them forward, typically using cables, to a local cable operator (LCO) who provides last mile connectivity to the home, also
using land lines. Recently, the LCO has also been fed with content
from satellites. This mode of cable TV is called Headend in the Sky
(HITS) due to the manner in which signals are transmitted to the
LCO. The different models of broadcasting and the associated issues
with respect to spectrum are discussed in subsequent sections.

Terrestrial TV: Public Airwaves

In the US, the FCC played a key role in regulating the delivery of
television programming. For several decades after World War II, the
FCC's television-related policies focused on conventional broadcasting, that is, the transmission of over-the-air television signals by local
TV stations to their surrounding communities. In fact, broadcasters
dominated the television market until the 1980s and are still characterized as trustees of their assigned spectrum blocks, obligated to
run programming 'in the public interest'. In a typical urban market,
FCC has licensed a handful of television stations to broadcast in the
VHF frequency band and a handful more to broadcast in the UHF
channels, whose over-the-air signals tend to travel far less due to
high frequencies and therefore reach fewer households. A VHF station is typically affiliated with one of the major television networks.
These networks include CBS (owned by Viacom), NBC (owned
by General Electric), ABC (owned by Disney), and Fox (owned by
News Corp.).

The FCC chairman Newton Minow's 1961 address to the National
Association of Broadcasters is legendary for its caustic dismissal of
television as 'a vast wasteland'. Yet Minow intended to emphasize a
different two-word phrase—public interest. Television was the most
prominent use of 'the people's airwaves,'—the government term used
to describe broadcast spectrum, clearly indicating its intended use
and it was failing to serve national interests. FCC rules effectively
limited the market to three major broadcast networks delivering
least-common-denominator content. The true wasteland was the
space where transmissions were not happening (Werbach 2009).

The same holds good for many other countries, including India.
Doordarshan (www.Ddindia.gov.in), today one of the largest

Table 11.6   Frequencies and number of channels provided by Doordarshan

Band	Spectrum	Number of TV channels available in analogue mode	TV channel number
UHF Band IV	470–582 MHz	14	21–34
UHF Band V	582–806 MHz	28	35–62

Source: Authors' own.

broadcasting organizations in the world in terms of the infrastructure of studios and transmitters, the variety of software, and the vastness of the viewership, had a modest beginning. Television and radio were both under the Ministry of Information and Broadcasting. Experimental telecasts started in Delhi in September 1959 with a small transmitter and a makeshift studio and regular daily transmissions started in 1965. The TV service was extended to a second city, Bombay, only in 1972. Till 1975, only seven cities were covered. Television was separated from radio in 1976 and Doordarshan came into existence. A national programme was introduced in 1982 and then onwards, there has been steady progress in Doordarshan.

Table 11.6 gives the frequencies and number of channels provided by Doordarshan through over-the-air broadcasting.

(1) UHF Band IV: There are 14 TV channels available in the UHF Band IV (470–582 MHz), each with 8 MHz channel bandwidth. Doordarshan operates around 330 transmitters in this band. Three digital TV transmitters in Kolkata, Chennai, and Mumbai are also being run on an experimental basis. Recently Doordarshan started its mobile TV service in Delhi using DVB-H technology on channel 26.

(2) UHF Band V: This band has 28 channels with 8 MHz bandwidth each from 582–806 MHz. No channel has been assigned for analogue TV transmission. However, frequencies 735–755 MHz and 775–795 MHz have been assigned in favour of Doordarshan to operate short distance UHF links for fixed line and data communications. Some of the government agencies are operating point to point microwave links in 610–806 MHz.

Presently, Doordarshan operates 35 channels—7 all-India channels (DD National, DD News, DD Sports, DD Gyandarshan, DD Bharti, DD Rajya Sabha, and DD Urdu), 11 Regional Language Satellite Channels (RLSC), 15 State Networks (SN), 1 international channel, DD India, and 1 HDTV channel. However, all these channels are also being provided through its free-to-air DTH service, 'DD—Direct Plus'.

Terrestrial TV offered by Doordarshan has lost appeal even among rural subscribers. Though about 92 per cent of the population in the country can receive Doordarshan programmes through a network of 1,415 terrestrial transmitters across the country, the substitutes—cable and DTH have taken over Doordarshan's strongholds, including its rural audience base. Thus terrestrial spectrum in India is also a 'vast wasteland' with possibilities for much better utilization and value creation.

Digital Dividend Spectrum

As indicated earlier, lower frequency radio waves have better propagation characteristics. Hence, the spectrum band in the 700 MHz range (698–806 MHz) in the UHF band V traditionally used for terrestrial television is much sought after by mobile operators. The wide reach offered by the 700 MHz spectrum necessitates fewer cell sites, thus reducing capital and operating expenses for the operators. It is approximately 70 per cent cheaper to provide mobile broadband coverage in the 698–806 MHz band than at 2100 MHz. This means networks can be rolled out quickly and cost-effectively, bringing cheaper services to consumers. Nearly 50 years after Minow stood before his stunned audience, the FCC has an opportunity, amid the vast wasteland of broadcasting, to create a verdant oasis of connectivity. However, governments need to find a way for broadcasters to vacate, either fully or partially, this precious spectrum space to make it available for high value services such as mobile.

The solution lies in 'digitization'. Once the broadcast signal is digitized, digital compression technologies and coding systems make it possible to squeeze much more information into a radio signal than in the case of analogue technology. The large amount of spectrum

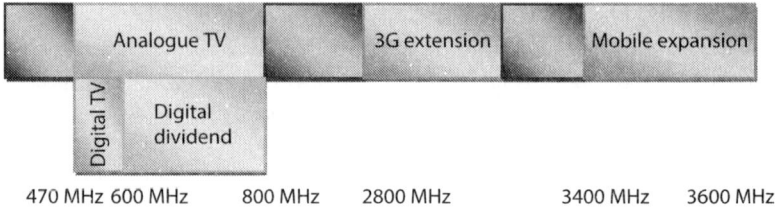

470 MHz 600 MHz   800 MHz   2800 MHz   3400 MHz   3600 MHz

Figure 11.2   Amount of spectrum that could be vacated by switchover from analog to digital

Source: Authors' own.

that would be freed up after the switchover from analogue to digital terrestrial TV is known as the 'digital dividend'.

Figure 11.2 illustrates the large proportion of spectrum, currently used for analogue TV, which can be vacated by the switchover from analogue to digital transmission. Most countries have planned for the digital switchover and some have already made the transition. Table 11.7 illustrates how the switchover is happening across the world.

As indicated in Table 11.7, the Nordic countries took the lead in digitization followed by other European countries and the US.

In India, cable digitization is in progress, and the due date for pan-India cable digitization is set for the 31 March 2015 (Kishore and Sridhar 2013). No date for the switchover from analogue to digital terrestrial transmission has been set although the digitization of the cable system may put pressure on the digitization of terrestrial transmission. Doordarshan is in the process of planning an upgrade of its 14 analogue TV channels transmitting in the range 470–582 MHz to digital TV transmission. The clauses that have been included in NFAP 2012, realizing the need for the allocation of digital dividend spectrum, are given in Table 11.8.

The released digital dividend spectrum is in the process of being allocated for commercial mobile services, especially for 4G services, worldwide. There are three major digital dividend plans: the European, the US and Asia Pacific Telecommunity (APT) band:

(1) European band in 790–882 MHz
(2) US band in 698–806 MHz
(3) APT band in 703–803 MHz

Table 11.7 Progress on digital switchover of terrestrial TV in select countries

Status	Countries (year of completion)
Transition is complete; all analogue signals terminated	Netherlands (2006), Finland (2007), Sweden (2007), Norway (2009), Switzerland (2007), Germany (2008), US (2009), Spain (2010), France (2011), South Africa (2011), UK (2012)
Transition in progress with expected date of complete switchover	Australia (2013), New Zealand (2013), Japan (2013), Colombia (2013), Chile (2013), South Korea (2013), Taiwan (2014), India (31 March 2015), Mexico (2015), Brazil (2016)
Transition no yet started; but planned	Most of the African countries except South Africa some of which including Rwanda, Namibia planning to switchover by 2015; China, Thailand, Venezuela
Not planned	Laos, North Korea

Source: Authors' own.

Table 11.8 Clauses in NFAP 2012 regarding digital dividend spectrum

IND 37	The requirement of digital broadcasting services including mobile TV may be considered in the frequency band 585–698 MHz subject to coordination on a case-to-case basis
IND 38	The requirement for IMT and broadband wireless access may be considered in the frequency band 698–806 MHz subject to coordination on a case-to-case basis

Source: WPC (2011).

The designated A5 and A6 bands as specified in the ITU recommendations and similar to the US band are given in Figure 11.3.

As illustrated earlier, there are two frequency channel arrangements/plans for the 698–806 MHz band. The first plan is based on Frequency Division Duplexing (FDD) and the second on Time Division Duplexing (TDD). In the FDD plan, there is 2×45 MHz bandwidth available from 703–748 MHz paired with 758–803 MHz

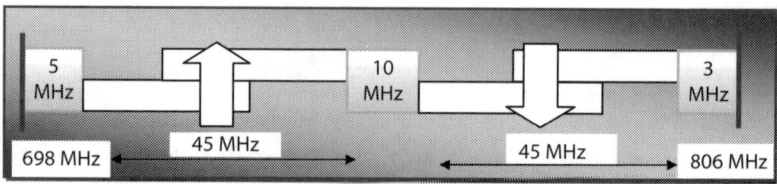

Harmonized FDD arrangement in the 698–806 MHz band

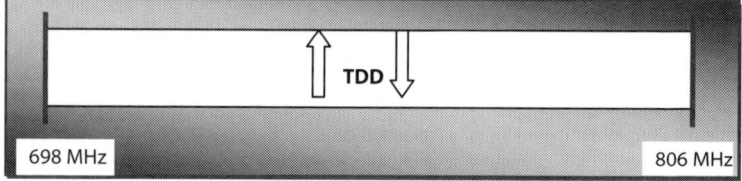

Harmonized TDD arrangement in the 698–806 MHz band

Figure 11.3   Harmonized FDD and TDD arrangement in 698–806 MHz band
Source: Authors' own.

with a 10 MHz centre gap and duplex separation of 55 MHz. Internal guard bands of 5 MHz (698–703 MHz) and 3 MHz (803–806 MHz) are provided at the lower and upper edges of the band for better coexistence with adjacent radio communication services. The TDD option of frequency arrangements in the band has also been considered with appropriate guard bands. In the TDD plan, a minimum internal guard band of 5 MHz at the lower edge (698 MHz) and 3 MHz at the upper edge (806 MHz) is proposed.

TRAI (2013) indicates the FDD plan is suitable for India and also provides a contiguous spectrum block of 2 × 45 MHz compared to the non-contiguous spectrum block as in the US plan. Contiguous spectrum enables operators to utilize the spectrum efficiently and hence is preferable. Moreover, the ecosystem for this band is stronger as most of the Asian countries and also Australia and New Zealand plan to allocate this band.

About 40 MHz of spectrum in this band has also been assigned to public sector units and government operators (BSNL and MTNL). However, TRAI (2012b) clearly indicates that the spectrum in the 696–806 MHz band is not being fully used and can be allotted for commercial mobile services. TRAI estimates that at least 2 × 30 MHz will be available for release.

Thus India, unlike most other countries, is in a unique position to take full advantage of the digital dividend spectrum. We have already noted the far greater commercial viability of 700 MHz as opposed to the 2100 MHz and 2300 MHz bands currently allocated for 3G and BWA services, respectively. It is crucial that the assignment of 700 MHz spectrum be through a well designed auction. TRAI has recommended that the auction of spectrum in 700 MHz band may be carried out preferably in 2014 as and when the ecosystem for LTE and LTE-Advanced is reasonably developed, so as to be able to realize the full market value of the spectrum. TRAI has also recommended fixing reserve prices at four times the reserve prices for the 1800 MHz band. This multiple may over-state the difference in the commercial viabilities of the respective bands and should be lowered in order to allow for a vibrant auction. One hopes that the digital dividend spectrum will be vacated by Prasar Bharati and allocated for commercial mobile services using a market based mechanism during 2014–15.

The migration to digital transmission in the US holds valuable lessons for all countries contemplating the move. In 2006, the United States Congress passed a resolution to switchover to digital television broadcasting by 2009. As of 1 March 2007, FCC required all television receivers shipped or imported into the US to have a digital tuner. In addition, effective 25 May 2007, the commission required sellers of television receiving equipment that did not include a digital tuner to disclose at the point-of-sale that such devices included only an analog tuner. In preparation for the switchover, most television broadcasters in the US simulcasted in both the analogue and digital formats. The FCC provided free bandwidth to broadcast in digital to existing analogue broadcasters. Most of the television sets in the US are capable of receiving digital signals via a built in digital tuners.

The FCC conducted auctions of part of the 'expected to be released' digital dividend spectrum in the 700 MHz band, officially known as Auction 73, on 24 January 2008 and the US government got a bounty of nearly $19.6 billion ($16.3 billion of which came from the two major operators Verizon Wireless and AT&T). The analogue to digital switchover occurred on 12 June 2009. Incidentally, in this auction apart from the telcos, Internet giant Google also participated, with a proposal to make the spectrum band 'open access' to all devices and applications to preserve net neutrality and allow leasing of the spectrum on a reasonable, fair, and non-discriminatory basis in the wholesale market to entities such as MVNOs. Google also guaranteed a minimum bid price of $4.6 billion with an upfront payment of $287,371,000. This in fact became the reserve price in the auction. Though Google did not eventually win the auction, FCC accepted the suggestion of open devices and open applications in this band and included the requisite clauses in the winners' licence guidelines.

Over the next 2 to 3 years, the FCC will conduct a series of auctions that could free up as much as 200 MHz of spectrum, depending upon how many TV broadcasters participate in the FCC's incentive auctions scheduled to begin in 2014. These auctions, called incentive auctions, seek to suitably reward broadcasters for freeing up frequencies while giving them enough flexibility to continue to broadcast, should they wish to do so, and clear visibility on the rules for the use of sold frequencies in order to assure against the threat of interference. Incentive auctions are at the heart of the FCC's mission to free up 300 MHz of spectrum for mobile broadband by 2015. That was the centrepiece of outgoing FCC chairman Julius Genachowski's mobile agenda and it will be the key in what his designated successor, Tom Wheeler, will need to deal with over the next several years. The FCC is currently writing the rules for the auctions and trying to persuade broadcasters to participate.

Incentive Auctions of 600 MHz in the US

The 2010 National Broadband Plan of the US introduced the idea of incentive auctions as a tool to help meet the nation's spectrum needs. Incentive auctions are a voluntary, market-based means of re-purposing spectrum by encouraging licensees to voluntarily relinquish spectrum usage rights in exchange for a share of the proceeds from an auction of new licences to use the re-purposed spectrum (FCC 2012). As per this scheme, broadcast television licensees holding spectrum in UHF bands (US channel numbers 14–36 in 470–608 MHz and channel numbers 38–51 in 614–698 MHz bands) may bid in the *reverse auction* to indicate the amount of compensation that they would accept to relinquish different spectrum usage rights, including the following (FCC 2012):

(1) 'all usage rights with respect to a particular television channel without receiving in return any usage rights with respect to another television channel';
(2) 'all usage rights with respect to [a UHF] television channel in return for receiving usage rights with respect to a [VHF] television channel (in the 54–72 MHz [low VHF]; 76–88 MHz [VHF]; and 174–216 MHz [high VHF] bands)'; or
(3) 'usage rights in order to share a television channel with another licensee'.

FCC floated this scheme to enable broadcasters to reap the benefits of flexible use (that is, to provide services other than broadcasting that include the mobile broadband service) in the digital transition, and also to help those who could not to take an alternate route to exit. Those who intend to exit would be compensated by an amount determined through the reverse auction. The exits would reduce the overall number of broadcast television stations competing for the same limited pool of advertising revenue and spectrum. Broadcasters who wish to remain in the broadcasting business also have an opportunity to strengthen their finances through the cash infusion resulting from a winning reverse auction bid to share channels or to move from a UHF to a VHF channel.

After the reverse auction, *re-packing* is carried out. This involves reorganizing the broadcast television bands so that the television stations that remain on the air following the auctions occupy a smaller portion of the UHF band, allowing the FCC to reconfigure a portion of the UHF band into contiguous blocks of spectrum suitable for flexible use. It is the necessity of re-packing that makes it necessary for the government to first conduct an incentive auction to free up spectrum and later a forward auction after reorganizing the freed spectrum. The mode of use of the freed spectrum is relevant to a broadcaster who wishes to stay in business. Hence the rules of the incentive auction are accompanied by a band plan for freed spectrum. Figure 11.4 illustrates such a band plan for 600 MHz. It depicts the spectrum blocks in sizes of 5 MHz that are suitable

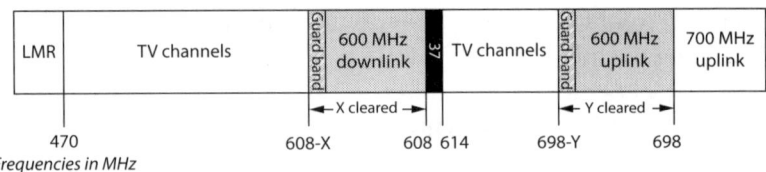

Figure 11.4   600 MHz band plan in the proposed incentive auction in the US
Source: FCC (2012).

for both broadcasting and advanced mobile wireless system released in the bands 608–X to 608 MHz and 698–Y to 698 MHz after the reverse auction; X and Y respectively refer to bands cleared through the reverse auction. The exercise of incentive auctions in the US, if successful, will release a maximum of 200 MHz for flexible use.

Satellite Television

International satellite television was introduced in India during 1991 with the coverage of the Gulf War by CNN. The signals were broadcast in the high frequency Ku band, received from INSAT and foreign satellites by the local cable TV operator at the head-end and transmitted through the local coaxial cable system to individual houses. Later the cable network value chain got disaggregated into an multi-service operator (MSO) and a Local Cable Operator (LCO). The MSO aggregated signals from broadcasters, encrypted them, and sent them to the LCO who further transmitted them to individual homes and managed customer relationships.

DTH television services referred to direct transmission of TV programmes from satellites to consumer-installed dish antennas using very high frequencies in the Ku band. Although transmitting in the Ku band and receiving programmes directly by users using dish antennas was prohibited until 2000 when DTH satellite TV guidelines were announced by the government, it was a common site in rural areas to see large antennas positioned to receive the C band transmission of television video signals.

While C-band antennas can vary between 6–8 feet due to the dispersion of low frequency radio signals, the high frequency Ku band is very focused and hence requires smaller dishes of about 18–24 inches, that can be easily deployed on roof tops.

In the US, although DTH providers first offered service in the 1980s, it was not until the 1990s that these providers assumed their current position as serious competitors to cable companies. There are two major DTH broadcasters in the US: DirecTV (started by Hughes Networks) and EchoStar.

The Government of India decided to permit DTH TV service using the Ku band in 2001. However, the initial adoption rate both by the service providers and subscribers was very low. It was only with the entry of new DTH operators in 2006 that subscriber growth started happening. In 2009, there were 10 million DTH subscribers distributed amongst five DTH service providers. This shot up to more than 40 million subscribers with the mandate of digitization of cable TV in the four metros by 31 October 2012. Since the digitization of cable signals would necessitate the presence of a digital tuner in the customer device, the cost of DTH and cable became qual. Most of the DTH operators are using the new MPEG4 standard. DVD quality picture and crystal clear sound quality make DTH services an option preferred to cable TV.

Competition in the DTH space is very high with about six–seven operators providing the service. One of the major hurdles that the DTH service providers face is the inadequacy of Ku band transponder space on the INSAT satellites.

Table 11.9 illustrates the major INSAT satellites that carry Ku band transponders for DTH services.

INSAT 4A satellites that host many of the DTH channels are nearing the end of their lives and hence the corresponding channels need to be shifted to other satellite transponders. Though many satellites with Ku band transponders are planned in 2013, they will not be able to meet the need for capacity. A shortage of Ku band transponder capacity is looming large on the horizon for DTH service providers. With an increasing number of TV channels and many standard definition channels getting converted into high definition ones, DTH players have to juggle with capacity allocations for their regional and national channels to retain their subscriber base.

As of now, there are over 800 satellite channels in India. At present, DTH players have anywhere between 12 and 15 transponders each (except Dish TV, which acquired six transponders a year-and-a-half ago to take its total to eighteen). While the best of compression

Table 11.9   INSAT satellites for DTH services

INSAT satellite	Transponders	Year of launch
4A	12-C; 12-Ku	Dec 2005
4B	12-C; 12-Ku	Mar 2007
4CR	12-Ku	Sep 2007
4D	24-C; 12-Extended C	Dec 2010; failed
4G	18 –Ku	May 2011
GSAT 10	18-C; 12 Ku	Sep 2012
GSAT 11	40 Ku and Ka band	Dec 2012
GSAT 14	6-C; 6-Ku	Feb 2013
GSAT 4R	6-C; 6-Ku	Dec 2013
GSAT 9	6-C; 24-Ku	Dec 2013
GSAT 14	6-C; 6-Extended C; 6 Ku; 2-Ka	Jan 2014

Source: Sridhar (2012).

technology can squeeze in 25 channels to a transponder, service providers who use MPEG 2 (compression format) can get far less. Hence, DTH operators have a capacity to transmit hardly 350 channels. This prevents DTH players from carrying many regional channels, although these channels are important for their subscriber base, particularly in rural markets.

As of now, DTH operators need at least an additional 10 transponders each to accommodate the demand for new channels. The DoT policy document states that while operations from Indian soil may be allowed to use both Indian and foreign satellites, proposals envisaging use of Indian satellites will be accorded preferential treatment (Business Line 2012).

The need for satellite space will only increase in the future as the number of TV channels is expected to double in 3 to 4 years. It is estimated that the DTH industry will need more than 220 transponders (the actual number of transponders in 2012 was 73) in 2017 to address the growth of DTH subscribers, the proliferation of television channels, and the provisioning of high density (HD) content. Under the government's Satellite Communication (SATCOM) policy, DTH operators are not permitted to directly buy or lease foreign transponders. These can only be procured by the Indian Space Research Organization (ISRO) on demand projection and sub-let to India-based users. Over 75 per cent of the 820 odd private channels are beaming into Indian homes through leases on foreign satellites. While ISRO tightly regulates this lease, it provides only 25 per cent of industry requirements on its own INSAT/GSAT communication satellites. It is also expected that more of INSAT and GSAT satellites with Ku band transponders will be put in to orbit to meet the growing demand. The Twelfth Plan working group on space noted that the space agency needs to 'pursue rigorously to secure spectrum' for another 100 Ku band and 50 C band/extended C band transponders. ISRO reckons that its 2017 tally should touch 400 transponders, including 102 in the C and 158 in Ku bands. But that will still fall below the last Plan target of 500 transponders, as well as the broadcasters' projected needs (Business Line 2013).

However, the future of satellite communication, especially for DTH operators, is in providing television channels bundled with broadband using high frequency spot beams in the Ka band (18–27 GHz). In India, this frequency has not yet been allocated and not even mentioned in NFAP 2011 with regard to DTH services. It is time that the satellite communication policy is revised to open up this band, which is extremely useful for providing HD TV as well as broadband connectivity by DTH service providers.

Table 11.10 gives the timeline of the policy and regulatory decisions in this sector.

Future Directions

As discussed in earlier chapter, non-exclusive dynamic use of TV white space (TVWS) and technologies such as super Wi-Fi to provide mobile communication services are being experimented with. West Virginia University in the rural hinterland in the US has started offering super Wi-Fi hot spots in its campus using TVWS.

A group of Microsoft researchers recently announced a complementary technical solution for unlicensed white spaces called WhiteFi (see Bahl et al. 2009; Naone 2009). WhiteFi is a complete system architecture for a Wi-Fi–like usage scenario operating within the white spaces. An access point can support multiple client devices

Table 11.10 Timelines of various policy and regulatory decisions in the television sector

Year	Event
1982	First INSAT satellite launched
1983	Doordarshan started cable TV service
1991	International satellite TV launched
1997	Prasar Bharati was formed; All India Radio and Doordarshan were brought under it
2001	DTH guidelines announced by DoT
31 Oct 2012	Mandatory digitization of signal to the home and implementation of the Conditional Access System in metros
12 Mar 2013	Government allows leasing of foreign transponder capacity by Indian DTH providers
31 Mar 2013	Second phase of cable TV digitization in 38 million plus cities
31 Mar 2015	Nationwide cable TV digitization to be completed

Source: Sridhar (2012).

such as laptop computers or mobile phones. WhiteFi networks could operate as local distribution points for regional-area connections to neighbourhoods.

In NTP 2012, there are clauses for promoting the use of white spaces with low power devices, without causing harmful interference to the licensed applications in specific frequency bands by deployment of software defined radios and cognitive radios. There are some initiatives in India by organizations such as the Centre for Internet Society for promoting the use of TVWs using the commons approach.

In 1999, as a precursor to TV white spaces, FCC chairman William Kennard embraced a modest initiative to allow low power FM (LPFM) broadcasters the right to broadcast in the unused guard band between the existing FM stations (Neuchterlein and Weiser 2005). This step, it was hoped, would give rise to a proliferation of stations devoted to community programming. The costs would be minimal, and the stringent caps on the transmission power of these 'micro radio' stations were such that their broadcast would not interfere with the existing FM stations to an unacceptable degree.

The National Association of Broadcasters vigorously opposed the move, citing the risks of interference posed by LPFM in the extreme case of a lone, immobile listener with low quality radio equipment. As a result LPFM was quashed for a number of years. Today, in the US, LPFM is allowed only for non-profit education and government agencies. However, countries such as New Zealand and Canada have adopted it more broadly. One of the important characteristics of these countries is the existence of remote areas where the LPFM can easily coexist with the extant high power FM broadcasting without causing harmful interference. This is an example of flexible spectrum management in the FM radio spectrum space, similar to the use of TV white spaces for Internet and commercial mobile services. It will be interesting to see if LPFM can provide broadband services in the future.

Conclusion

Entertainment is a big part of the daily lives of many people in India, especially in suburban and rural areas. FM Radio, cable TV, and DTH provide services in this domain. Due to convergence, it is possible for all operators to provide advanced communication services including Internet access. Hence, the demand for spectrum in the associated bands is expected to increase and it is vital that the assignment of such spectrum is managed efficiently. In this chapter we looked at the various methods of assignment of spectrum for broadcast, and also explored newer areas of development within the digital dividend spectrum and TV white spaces that can be allocated for alternative uses including mobile and Internet access.

References

Bahl, P., Chandra, R., Moscibroda, T., Murty, R., and Welsh, M. (2009). White space networking with wi-fi like connectivity. *ACM SIGCOMM Computer Communication Review* 39(4): 27–38.

Business Line. (2012). 'DTH players clamour for more bandwidth', 26 July.

———. (2013). 'Beams us up, Bangalore', 20 April.

Federal Communications Commission (FCC). (2012). 'Notice to the proposed rulemaking: FCC 12–118.' Available at: www.fcc.gov. Accessed on 8 December 2012.

Hazlett, T. (1990). 'The rationality of U.S. regulation of the broadcast spectrum', *Journal of Law & Economics*: 133–74.

Kishore, K., and V. Sridhar. (2013). 'A potential solution to cable digitization.' *Financial Express*, 7 March.

Naone, E. (2009). *Wi-Fi via White Spaces*, TECH. REV, 18 August. Available at: http://www.technologyreview.com/communications/23271/?a=f. Accessed on 16 October 2013.

Neuchterlein, J., and P.J. Weiser. (2005). *Digital crossroads*. Cambridge, MA: MIT Press.

Sridhar, V. (2012). *Telecom Revolution in India: Technology, Regulation and Policy*. New Delhi: Oxford University Press.

Telecommunications Regulatory Authority of India (TRAI). (2008b). 'Consultation paper on issues relating to 3rd phase of private FM radio broadcasting', January. Available at: http://www.trai.gov.in. Accessed on 15 January 2011.

————. (2012a). 'Recommendations on auction of spectrum.' Available at: www.trai.gov.in. Accessed on 10 February 2013.

————. (2012b). 'Recommendations on "Prescribing minimal channel spacing within licensed service area in FM Radio sector in India.' Available at: www.trai.gov.in. Accessed on 8 January 2013.

————. (2013b). 'Recommendations on IMT-Advanced mobile wireless broadband services.' Available at: www.trai.gov.in. Accessed on 25 March 2013.

Werbach, K. (2009). 'The wasteland: Anticommons, white spaces and the fallacy of spectrum', *Arizona Law Review* 53(1): 213–54.

Wireless Planning and Coordination Wing (WPC). (2011). *Draft India Remarks in the National Frequency Allocation Table*. Available at: http://www.dot.gov.in. Accessed on: 10 January 2013.

12

The Way Forward

We will be running an LTE signal and understanding the impact of radar on that LTE signal. Hopefully, no missile will be fired as part of the test.

—Jeff Reed, Director, Wireless @ Virginia Tech

Spectrum as an excludable and partially rivalrous economic resource is a toll good that is amenable in the face of new technological developments and exploding demand to an increasing amount of non-exclusive use. At the same time structures for the private use of spectrum, particularly in dense urban agglomerations, continue to remain relevant. The management of spectrum in the face of these diverse pressures and in view of the immense benefits that could accrue from optimal use, promises to become a case study leading to continued technological development, commercial innovation, and regulatory breakthrough.

Progress, as always, comes partly by design, partly by the force of necessity, and often by sheer serendipity. The immense pressure on wireless networks, that is only expected to grow in the face of '50 billion' devices coming online, is forcing market players to accelerate technological progress and to come up with new ways of spectrum management including non-exclusive use. The outcomes of experiments in the private sphere including spectrum sharing and Femtocells, form a rich source of proof of concept on futuristic models of spectrum management. The phenomenon of network externalities holds out the prospect of exponential growth as well as expedited extinction, and is forcing companies to come together in innovative partnerships to create tightly coupled value networks that engage in extreme competition with each other.

We conclude in this chapter our arguments for a path towards dynamic and flexible management of spectrum for providing universal access to the Internet and associated applications on mobiles for socioeconomic development.

Legacies

As described in chapters 6 and 10, India is one of the many emerging countries where spectrum is doled out in small chunks, often at high prices, resulting in excessive spectrum fragmentation. A large proportion of spectrum is directly in the control of the government, making the public sector an unavoidable bottleneck, or a crucial ally, as the case may be.

The legacy of spectrum management forms the base from which new models must emerge. This legacy was forged in times when wireless technology was far less advanced, commercial spectrum was extremely scarce, and the role of mobile devices in society only peripheral to the role of fixed line telephony and Internet. The realities of those times led to specific kinds of licences including bundling of the spectrum and service licences, mandates on technology and service, strictures on secondary market activity, and obligations for a certain pace of roll-out and rural connectivity. Dismantling the anachronistic superstructure in the face of increased amounts of spectrum availability, advanced technologies, greater possibilities of non-exclusive use, and the promise of mobile devices being the via media for connecting large parts of the globe to the Internet, is an absolute necessity. However, the task must proceed in a phased manner keeping in mind political imperatives, the role played by revenues from the ICT sector in public finance, and the sensitive role of media in a society.

It is time that the government, policymakers, and regulators depart from the legacy and provide an environment for flexible spectrum management. The rationalization of the use of government spectrum, and the freeing up of spectrum for private and shared use, form the tasks of the government as a user of spectrum, in addition to its role as a facilitator of spectrum markets and sharing. Indeed, well-functioning spectrum markets are necessary for the government to give up some of its spectrum (to provide a channel for the acquisition of spectrum in case of future need), but such markets can only come into being provided the government frees up sufficient quantities of spectrum. In a rare move, the US Navy allowed the use of radio frequencies allotted for radar navigation system to be simultaneously used for testing LTE networks in nearby locations (IJEC 2013).

As technology evolves at Internet speed, it is time that spectrum policies are forged in the spirit of technology and service neutrality to provide space for future evolution.

Technologies

Technologies in the mobile space continue to evolve at an astonishing pace as outlined in Chapter 1. In a rare move, Facebook, the Internet company, has taken the lead along with the mobile device makers such as Nokia and Samsung, chip makers such as Qualcomm, and network equipment makers such as Ericsson to create Internet.org—an organization whose mission is to provide Internet connectivity on mobile universally to everyone, particularly to those in the developing world (NYT 2013).

Exploiting the white space spectrum, augmenting capabilities of end-user terminals with cognitive radio capabilities in end-user devices, deploying storage solutions that store mobile content nearer to the users in a distributed manner, implementing effective in-building solutions for improving quality of service, effectively integrating wired and wireless networks, supplementing carrier networks with unlicenced wireless LAN networks, and building intelligent and context-aware mobile applications are some of the technology trends that we have delved upon in different chapters in this book. The challenge is to make these technologies available at affordable prices and drive economies of use, especially as appropriate for the needs of emerging countries. One hopes that innovative partnerships and consortia, such as the one discussed here and others discussed in Chapter 9 will be able to address the present challenge.

Economics

Finally, it is economics that must underpin the long term sustainability of any initiative. As discussed in chapters 2 to 5, the economics of spectrum is quite complex and enjoins upon us to carefully choose and design allocation and assignment mechanisms while not losing sight of legacy issues, technology dynamism, price sensitivity, and the imperatives of competition. Cases of spectrum being hoarded, or landing up in wrong hands can be minimized by a greater focus on recent advances in the design of auctions, pragmatic methodologies on estimating spectrum value, and a clear understanding of the tradeoff between present revenue from spectrum and future revenues from services.

Conclusion

The growing virtualization of the world, powered by the Internet, is forcing us to re-examine the Internet in the light of the role played by different sectors like media, health, education, and entertainment in our societies. At the same time, these sectors are being redefined in the context of the possibilities of the Internet. A good example of this creative tension lies in the sphere of media where old models of regulation of news focusing on limiting the market power of content providers are coming face to face with new paradigms of regulation of the Internet that emphasize absolute freedom for content providers and regulation of the intermediary 'pipes' transporting content to the end users, as discussed in Chapter 8.

Debates on these complex issues is particularly urgent in developing economies like India where wireless connectivity holds out the promise of being a substitute for older modes of access and hence an opportunity to leapfrog unrealized goals on traditional infrastructure. While the potential of the Internet cannot be overstressed, one must not forget the complementarity of power, the need for developing requisite skills, and the imperative of building applications that will draw masses to the new technology.

The journey to a connected world has just begun and requires us to realign our governance structures and economies around the new technology, while remaining alive to the dangers inherent in information overload, over-connectivity, invasion of privacy, and new avenues of security risks. Creating this brave new world is the happy task of new generations of scientists, engineers, entrepreneurs, investors, and regulators. The need of the hour is for different mindsets and diverse skill sets to come together, to integrate perspectives, and to synthesize viewpoints for the new world cannot be built in silos and dispensed to customers over the counter.

This book represents a modest attempt at bringing together technological and economic perspectives in the sphere of spectrum management, keeping legacy issues squarely in focus when making recommendations. The management of spectrum is dynamic in two respects—dynamic in the sense of ever-changing in response to new technologies and business models, and dynamic in the sense of the technological feasibility of dynamically assigning frequencies to users without having to apportion spectrum to private use. To the skillful management of these inherent dynamisms the book is dedicated.

References

International Journal of Electromagnetic Compatibility (IJEC). (2013). 'US Navy tests military spectrum sharing potential.' Available at: http://www.interferencetechnology.com/u-s-navy-tests-military-spectrum-sharing-potential/. Accessed on 10 August 2013.

New York Times (NYT). (2013). 'Facebook leads an effort to lower barriers to Internet access', 21 August.

(Continued)

Chapter	Term	Description
1	3GPP	Third Generation Partnership Project: The industry group that oversees the creation of industry standards for the third generation of mobile wireless communication systems. The key members of the 3GPP include standards agencies from Japan, Europe, Korea, China, and the United States. More information about 3GPP can be found at www.3GPP.org
1	4G	Fourth Generation: Mobile telecommunication systems offering significantly high speed data and multimedia services (typically 100 Mbps or more). LTE is the widely used 4G technology around the world.
1	CDMA	Code Division Multiple Access: The multiple access scheme developed by Qualcomm, USA to support multiple simultaneous communications over the wideband frequency by adding a unique code for each data signal that is being sent to and from each of the radio transceivers. CDMA development group is a consortium of companies that promote the evolution of CDMA. More information can be found at www.cdg.org
1	Cognitive radio	An intelligent radio that can be programmed dynamically to send and receive radio signals depending on radio channels available in the vicinity.
1	FCC	Federal Communications Commission: The independent regulatory and licensing authority for the telecommunications and broadcasting sector in the USA, set up by the Communications Act of 1934.

Glossary*

Chapter	Term	Description
1	2G	Second Generation: Digital mobile telecommunication systems that provides basic voice and minimum speed data connectivity. It uses circuit switching for voice and packet switching for data communication. GSM is the widely used standard in 2G while CDMA is a proprietary standard developed by Qualcomm, USA.
1	3G	Third Generation: Digital mobile telecommunication systems that provide high-speed data connectivity and enable video calling, mobile TV, and other multimedia services apart from supporting high-quality voice service. The specifications for 3G were set up by the International Mobile Telecommunications-2000 IMT 2000 forum of the ITU. The 3GPP provides standardization support for 3G services. WCDMA, CDMA-EVDO, HSPA, HSDPA, HSUPA are examples of technologies that come under the umbrella of 3G services.

(Continued)

* Terms have been arranged chapterwise, alphabetically.

FDMA	1	Frequency Division Multiple Access: Process of allowing mobile radios to share radio frequency allocation by dividing up that allocation into separate radio channels where each radio device can communicate on a single radio channel during communication.
GSM	1	Groupe Speciale Mobile (Global Systems for Mobile): A public all-digital cellular network standard using the transmission band around 900 MHz, 1800 MHz, and 1900 MHz developed by the European Telecommunication Standards Institute (ETSI). A GSM network can provide, besides telephony services, Short Messaging Services (SMS) and data communication, in circuit and/or packet mode. GSM Association represents the corresponding interest group. More information is available at www.gsm.org
GSMA	1	GSM Association is the industry body formed in 1995 to promote the development, standardization, and adoption of GSM and associated technologies.
HSDPA	1	High Speed Downlink Packet Access: High-speed 3G digital data services based on WCDMA with a downlink speed of 14.4 Mbps and uplink speed of 400 Kbps.
HSPA	1	High Speed Packet Access: A family of high-speed 3G digital data services based on WCDMA, with a downlink speed of 42 Mbps and uplink speed of 5.8 Mbps.
HSUPA	1	High Speed Uplink Packet Access: High speed 3G digital data services based on WCDMA with a downlink speed of 14.4 Mbps and uplink speed of 5.7 Mbps.

(Continued)

(Continued)

Chapter	Term	Description
1	IEEE	Institute of Electrical and Electronics Engineers: The world's largest professional association dedicated to advancing technological innovation and excellence for the benefit of humanity. More information can be found at www.ieee.org
1	IMT 2000	The International Mobile Telecommunications standard developed by the International Telecommunications Union for the advancement of mobile communications in the twenty-first century.
1	IMT-MC/DS	One of the CDMA-based 3G standard promoted by ITU under IMT 2000 standing for Multi Carrier/Direct Spread widely adopted by Japan.
1	ITU	International Telecommunications Union: ITU is the United Nations specialized agency for information and communication technologies (ICTs). ITU allocates global radio spectrum and satellite orbits, develops the technical standards that ensure networks and technologies seamlessly interconnect, and strive to improve access to ICTs to underserved communities worldwide. More information can be found at www.itu.int
1	LTE	Long Term Evolution: The next generation 4G technology for both GSM and CDMA cellular carriers. Approved in 2008 by 3GPP with download speeds up to 173 Mbps.
	OFDMA	Orthogonal Frequency Division Multiple Access that enables large data tranmission by splitting the information into Multiple smaller chunks that are then simultaneously transmitted at different carrier frequencies orthogonal to each other.

TDMA	1	Time Division Multiple Access: Process of allowing mobile radios to share radio frequency allocation by dividing up the time slots that are shared between simultaneous users of the radio channel. When a mobile radio communicates with a TDMA system, it is assigned a specific time position on the radio channel.
TD-SCDMA	1	Time Division-Synchronous Code Division Multiple Access: High-speed 3G mobile communication standard that combines support for both circuit-switched and packet-switched data. The standard combines time division multiple access (TDMA) with an adaptive, synchronous-mode code division multiple access (CDMA) component. TD-SCDMA was developed by the China Academy of Telecommunications Technology (CATT) in collaboration with Datang and Siemens.
UHF	1	Ultra High Frequency: Designates the ITU radio frequency range of electromagnetic waves between 300 MHz and 3 GHz (3,000 MHz), also known as the decimeter band or decimeter wave as the wavelengths range from 1 to 10 decimeters (10 cm to 1 m).
UMTS	1	Universal Mobile Telecommunications System: Set of wireless communication standards that enable the use of a combination of wireless and fixed systems in an effort to provide seamless telecommunications services to its users. The specifications for the UMTS system are overseen by the 3GPP.
VHF	1	Very High Frequency: The radio frequency range from 30 MHz to 300 MHz traditionally used for FM radio broadcasting.

(Continued)

Chapter	Term	Description
	VoIP	Voice Over Internet Protocol—Process of sending voice traffic in the form of packets over IP networks. VoIP digitizes analog voice, compresses it, packetizes and sends it to the receiver.
1	WCDMA	Wideband Code Division Multiple Access: Wideband code division multiple access is a third generation mobile communication system that uses CDMA technology over a wide frequency band to provide high-speed multimedia and efficient voice services. The WCDMA infrastructure is compatible with GSM mobile radio communication system.
1	Wi-Fi	Wireless Fidelity: Local area wireless networks based on IEEE 802.11 b/g/n standards operating in the licence-free Industrial Medical and Scientific (ISM) radio frequency bands of 2.4 GHz and 5 GHz. Internet services using this technology are normally provided by the ISPs.
2	Coase Theorem	If the parties involved can costlessly communicate with each other, then they can arrive at an efficient solution to the problem of externalities.
2	Common property resources	Such resources are non-excludable but rivalrous. For example, a common pasture.
2	Commons goods	Commons goods are non-excludable and rivalrous.
2	Excludable	A good is referred to as excludable if it is possible to exclude users from its use.
2	Externalities	Uncompensated impact of one economic actor on another. For example, in passive smoking.

2	Managed commons	Shared use of unlicensed spectrum with certain rules of the game specified.
2	Marginal cost	Increase in total cost for each additional good.
2	Marginal revenue	Increase in total revenue for each additional good.
2	Open access	A regime under which anyone has access to an unowned resource without limitation.
2	Private commons	Shared use of licensed spectrum with certain rules of the game specified.
2	Private goods	Private goods are both excludable and rivalrous.
2	Pure public goods	Pure public goods are non-excludable and non-rivalrous.
2	Rivalrous	If the use of a good by a user precludes its consumption by another user, then the good is said to be rivalrous.
2	Toll goods	Toll goods are excludable and non-rivalrous.
2	Tragedy of the anti-commons	The process of innovation becomes cumbersome, unwieldy, and inefficient because of the large number of patents that must be licensed.
2	Tragedy of the commons	A partially rivalrous good treated as a common property resource is subject to congestion, as a result of the uncoordinated pursuit of self-interest.
3	Contestability	An industry is said to be perfectly contestable if three conditions are satisfied: firstly, new firms face no disadvantage with respect to existing firms; secondly, there are zero sunk costs; thirdly, the entry lag is less than the price adjustment lag.
3	Economies of scale	Reduction in per unit costs with an increase in the scale of operation.

(Continued)

(Continued)

Chapter	Term	Description
3	Economies of scope	Reduction in per unit costs of a line of production through the operation of related lines of production.
3	Flexible use	The job of spectrum managers would end at carving out blocks and then auctioning them with rules to prevent interference.
3	Mobile number portability	Allowing a subscriber to retain the mobile number even after changing service providers.
3	Network externalities	The beneficial impact of the number of users on the benefits enjoyed by each user.
4	Ascending auction	Bidding starts from a low amount and moves up.
4	Beauty contests	A process of administrative hearings to choose individual licensees of spectrum.
4	Collusion	Illegal cooperation between market players to subvert competition.
4	Common value auctions	All bidders have the same value for the object but different signals/information about it, and therefore different estimates.
4	Descending auction	The auctioneer sets a very high price and continuously lowers it till a bidder calls out to buy.
4	Dutch auctions	The auctioneer announces a very high price and progressively lowers it until a bidder indicates that she is prepared to buy. She is declared the winner and pays the bid price.
4	Dynamic auction	Bidders can observe the bids of other.

	Term	Description
4	English auction	Open ascending-bid auctions in which the auction starts at a low price and moves up with bidders dropping out as the price becomes too high for them. The price is raised by 'open outcry' of the bidders.
4	First-price auction	In a first-price auction, the highest bidder gets the object at a price equal to their bid amount.
4	Japanese or 'clock' auction	Open ascending-bid auctions in which the auction starts at a low price and moves up with bidders dropping out as the price becomes too high for them. The price is raised in predetermined increments by the auctioneer.
4	Multi-unit auction	More than one object is auctioned.
4	NPV	Net Present Value, the present value of a future stream of income, using a discount rate like the prevailing rate of interest.
4	Package auction	Multi-unit auctions held simultaneously; bidders bid for combinations of objects.
4	Private value auctions	(a) Each bidder has a personal value for the object which is independent of the value of other bidders and (b) each bidder knows the value the object has for her.
4	Reserve price	A reserve price is the price at which bidding begins. The auctioneer will not accept a lower bid.
4	Second-price auction	In a second-price auction, the highest bidder gets the object at a price equal to the bid of the second highest bidder.
4	Simultaneous ascending auction (SAA)	Multi-unit auctions held simultaneously; bidders bid separately for each object.

(Continued)

Chapter	Term	Description
4	Static/sealed bid auction	Bidders make a single bid without observing the bids of other bidders.
4	Vickrey auctions	Second-price sealed-bid auctions.
4	Winner's curse	A phenomenon where the winner of an auction pays more than the value of the object.
5	AGR	Adjusted Gross Revenue: The gross revenue accruing to the telecom licensees by way of operations of the celluar mobile service and also includes
5	AIP	Administered Incentive Pricing: A method of valuing spectrum by its opportunity cost.
5	ARPU	Average Revenue per User
5	Cash flow	The annual cash accrual from the possession of a given block of spectrum.
5	Category A	The second highest LSAs in terms of revenue per user.
5	Category B	The third highest LSAs in terms of revenue per user.
5	Category C	The fourth highest LSAs in terms of revenue per user.
5	Cobb-Douglas production function	A specific kind of production function.
5	Intrinsic value	Valuation of a band of spectrum based only on its propagation capability.
5	Licence fee	A percentage of adjusted gross revenue taken from operators towards the cost of the service licence.

5	LSA	Licensed Service Area: Geographically demarcated telecom market in India, also called 'circle'.
5	Metro	The highest LSAs in terms of revenue per user.
5	MP	Marginal productivity, the increment in output for a unit increase in a particular input, for example, labour.
5	Opportunity cost	The cost of inputs saved by virtue of possessing an additional megahertz of spectrum.
5	Production function	A mathematical function depicting the relationship between inputs and outputs in a production process.
5	Spectrum charge	A percentage of AGR taken from operators towards the cost of spectrum.
6	BWA	Broadband Wireless Access: An access service that provides high-speed data connectivity through mobile devices.
6	Command and control	Spectrum management regime in which many of the following decisions are administratively undertaken: spectrum use, choice of technology, number of blocks, the identification of the individual users, and the price paid.
6	DoT	Department of Telecommunications: Department under the Ministry of Communications and IT of the Government of India that oversees the telecommunications sector of the country including policymaking, spectrum allocation, and licensing. More information can be found at www.dot.gov.in
6	Femtocells	Femtocells are small micro-pico-cells that are deployed as In-Building Solutions operating in licensed bands that allow traffic to be diverted from the carrier's macro cellular network to a localized network, thus relieving the load on the licensed spectrum of the macro network.

(Continued)

Chapter	Term	Description
6	INSAT	Indian National SATellite: A system of satellites that are placed in geostationary orbits in the Asia-Pacific region by ISRO. More information is available at http://www.isro.org/satellites/geostationary.aspx
6	MNO	Mobile Network Operators: A licensed telecom operator who has spectrum and other associated facilities to provide mobile communication services.
6	MVNO	Mobile Virtual Network Operator: A licensed/registered telecom operator who normally leases/rents spectrum and other associated facilities to provide mobile communication services.
6	NFAP	National Frequency Allocation Plan prepared by the Wireless Planning and Coordination Wing of the Department of Telecommunications periodically to designate different spectrum bands for different usages.
6	USO	Universal Service Obligation: The telecom policy that enables the availability of telecommunication services in a non-discriminatory manner to all citizens of the country. Normally funded by USO Fund, collected through USO Levy from the telecom operators.
6	WiMAX	Worldwide interoperability for Microwave Access: Technology and protocols envisioned and supported by WiMAX forum, later adopted by IEEE as 802.16 standard for high-speed fixed and mobile wireless service, typically in spectrum blocks in 2 GHz to 11 GHz range. More information can be found at www.wimaxforum.org

6	WPC	Wireless Planning and Coordination wing; The unit in the DoT responsible for radio frequency spectrum planning and allocation.
7	DLNA	Digital Living Network Alliance is a set of protocols advocated by the DLNA group to connect various devices at home including wireless speakers, television, set top boxes, mobiles, tablets, and PCs over home Wi-Fi networks to share images, music files, and videos.
7	DSA	Dynamic Spectrum Access is a set of technologies protocols for dynamically sharing spectrum among different networks.
7	OTT	Over-The-Top (OTT) players: Internet companies such as Skype and Whatsapp which provide Internet-based services such as VoIp and messaging over the data network connectivity of mobile carriers. The carriers do not have control over what is transmitted by the OTT players to the end users and act as just bandwidth pipe providers.
7	Mi-Fi	Technology with this trade name that allows WiFi local access to a number of mobile devices, with a 3G or 4G technology for backhaul; typically can provide connectivity to multiple devices on the move.
7	RFID	Radio Frequency Identification: Technology that uses a small chip or tag attached to products, animals, or vehicles embedded in cards for the purposes of identification or tracking.

(Continued)

(Continued)

Chapter	Term	Description
7	SIM	Subscriber Identification Module: It is an integrated circuit that securely stores the International Mobile Subscriber Identity (IMSI) used to identify a subscriber on mobile telephony devices. A SIM is held on a removable SIM card often provided by the mobile operator, which can be transferred between different mobile devices. A SIM card contains its unique serial number known as Integrated Circuit Card Identifier (ICCID), security authentication and ciphering information, temporary information related to the local network, a list of the services the user has access to, and passwords.
8	CAP	Content and Application Providers who provide content and application to the end users, typically through landline or mobile network services.
8	ECP	End-user Connectivity Provider who provides the last-mile access to end users; typically mobile and landline operators belong to this class.
8	ISP	Internet Service Providers who provide Internet service through their own infrastructure or leased infrastructure.
8	Net Neutrality	The principle that ISPs treat all content and applications equally, and shall not discriminate in terms of priority, speed, and other related mechanisms.

10	CMSP	Cellular Mobile Service Providers who have been licensed and assigned spectrum for providing cellular mobile services.
	HHI	Herfindashl–Hirschman Index (HHI): Measure of the size of firms in relation to the industry and an indicator of the amount of competition among them. It is calculated as $\sum si^2$ Where s_i is the share of the firm 1 (for example, subscribers) in the market. A value of 1 indicates monopoly, while 0.18 indicates perfect competition.
10	SBC	Subscriber Based Criterion that determines percentage of revenue the mobile operator needs to pay based on spectrum holding.
10	TRAI	Telecom Regulatory Authority of India: The regulatory organization created on 28 March 1997 under The Telecom Regulatory Authority of India Act 1997 to regulate telecommunication services in the country. More information can be found at www.trai.gov.in
10	UASL	Unified Access Service Licence: Type of access service licence formulated in 2003 by the Government of India to provide any service using any technology.
10	UASP	Unified Access Service Provider: Telecom operator who has UASL.
11	AM broadcasting	Public over-the-air radio broadcasting using amplitude modulation. The AM radio operates at much lower frequencies and hence propagates over a larger geographical area compared to FM radio.

(Continued)

Chapter	Term	Description
11	DTH	Direct To Home: Television broadcasting service that uses Ku and associated high-frequency radio frequencies to transmit video programmes from geostationary satellites to small antennas mounted on the rooftops of households.
11	FM broadcasting	Public over-the-air radio broadcasting using frequency modulation. The FM radio can be received through an antenna normally provided in devices such as mobile handsets and car radio units, typically in 88 MHz–108 MHz.
11	TVWS	Television White Space that refers to small blocks of spectrum in between adjacent television channels that are owned by the television broadcaster; however not used most of the time; can potentially be used by other users.
11	EV-DO	Evolution Data Optimized: A 3G digital service provided by CDMA cellular carriers. Part of the CDMA2000 standards, EV-DO system provides for multiple voice channels and medium-rate data services.

Bibliography

Aggarwal, V. (2006). 'Jagadish Chandra Bose: The real inventor of Marconi's wireless detector', *Ancient Wireless Association Journal* 47 (3, July): 50–54.

———. (2013). Sir Jagadish Chandra Bose. Available at: http://web.mit.edu/varun_ag/www/bose.html. Accessed on 10 October 2013.

Akhtar, S. (2010). '2G-4G Networks: Evolution of technologies, standards, and deployment', *Encyclopedia of Multimedia Technology and Networking*. Ideas Group Publisher. Available at http://faculty.uaeu.ac.ae/s.akhtar/EncyPaper04.pdf. Accessed on 7 April 2013.

Alden, J. (2012). 'Exploring the value and economic valuation of spectrum.' Report prepared for the ITU. Available at: http://www.itu.int/ITU-D/treg/broadband/ITU-BB-Reports_SpectrumValue.pdf. Accessed on 31 October.

Andrews, J., H. Claussen, M. Dohler, S. Rangan, and M. Reed. (2011). 'Femtocells: Past, present, and future', *IEEE Journal on Selected Areas in Communications* 30(3, April): 497–508.

Bahl, P., R. Chandra, T. Moscibroda, R. Murty, and M. Welsh. (2009). 'White space networking with Wi-Fi like connectivity', *ACM SIGCOMM Computer Communication Review* 39(4): 27–38.

Baumol, W. (1982). 'Contestable markets: An uprising in the theory of industry structure', *American Economic Review* 72(1): 1–15.

BEREC. (2012a). *BEREC Report on differentiation practices and related competition issues in the scope of net neutrality*. Available at: http://berec.europa.eu/eng/document_register/subject_matter/berec/reports/1094-berec-report-on-differentiation-practices-and-related-competition-issues-in-the-scope-of-net-neutrality. Accessed on 10 March 2014.

———. (2012b). 'Report on differentiation practices and related competition issues in the scope of net neutrality.' Available at: http://berec.europa.eu/eng/document_register/subject_matter/berec/reports/?doc=1094. Accessed on 8 August 2013.

Bose, J.C. (1899). 'On a self-recovering coherer and the study of the cohering action of different metals', *Proceedings of the Royal Society* LXV(416): 166–172. Reprinted in January 1998, *IEEE Proceedings* 86(1): 244–7.

Brain, Marshall. (2000). 'How the radio spectrum works.' Available at: http://electronics.howstuffworks.com/radio-spectrum.htm. Accessed on 18 December 2012.

Brandenburger, Adam M., and Barry J. Nalebuff. (2011). *Co-opetition*. NY, USA: Crown Business.

Breeden, John. (2012). 'Is the light spectrum next frontier for wireless?' 6 December. Available at: http://gcn.com/blogs/emerging-tech/2012/12/light-spectrum-next-wireless-frontier.aspx. Accessed on 11 December 2012.

Bullington, K. (1953). 'Frequency economy in mobile radio bands', *Bell System Technical Journal*: 42–62. Available at: http://www.alcatel-lucent.com/bstj/vol32-1953/articles/bstj32-1-42.pdf. Accessed on 7 November 2012.

Brito, J. (2006). 'The spectrum commons in theory and practice.' *Stanford Technology Law Review*. Available at: http://stlr.stanford.edu/pdf/brito-commons.pdf. Accessed on: 15 February 2013.

Business Line. (2012). 'DTH players clamour for more bandwidth', 26 July, *Hindu Business Line*.

―――. (2013). 'Beams us up, Bangalore', 20 April, *Hindu Business Line*.

Federal Communications Commission (FCC). (2012). 'Notice to the proposed rulemaking: FCC 12–118.' Available at: www.fcc.gov

Bykowski, M. (2003). 'A secondary market for the trading of spectrum: Promoting market liquidity', *Telecommunications Policy* 27: 533–41.

Carter, Kenneth, Ahmed Lahjouji, and Neal McNeil. (2003). Unlicensed and unshackled: A joint OSP-OET white paper on unlicensed devices and their regulatory issues.' Available at: http://www.fcc.gov/working-papers/unlicensed-and-unshackled-joint-osp-oet-white-paper-unlicensed-devices-and-their-regu

Caicedo, C.E., and M.B. Weiss. (2011). 'The viability of spectrum trading markets', *IEEE Communications Magazine* 49(3): 46–52.

Cave, M., C. Doyle, and W. Webb (2011). *Essentials of Modern Spectrum Management*. Cambridge, UK: Cambridge University Press.

Chapin, J., and W. Lehr (2007). 'The path to market success for dynamic spectrum access technology.' *IEEE Communications Magazine* 96–103.

Cheng, H.K., S. Bandyopadhyay, and H. Guo. (2011). 'The debate on net neutrality: A policy perspective', *Information Systems Research* 22(1): 60–82.

Choi, J.P., and B.-C. Kim. (2010). 'Net neutrality and investment incentives', *RAND Journal of Economics* 41(3): 446–71.

Cisco. (2010). 'Cisco visual networking index: Global mobile data traffic forecast update, 2010–2015.' Available at: http:::/www.cisco.com

Coase, Ronald H. (1959). 'The federal communications commission', *Journal of Law and Economics* 2 (October): 1–40.

Coase, Ronald Harry. (October, 1960). 'The problem of social cost', *Journal of Law & Economics* III: 1–44.

Cramton, Peter, Yoav Shoham, and Richard Steinberg. (2006). 'Simultaneous ascending auctions', 99–114.

Crocioni, P. (2009). 'Is allowing trading enough? Making secondary market in spectrum work', *Telecommunications Policy* 33: 451–468, doi:10.1016/j.telpol.2009.03.007.

Danigelis, A. (11 April 2012). 'Wireless could have saved lives on the *Titanic*.' Available at: http://news.discovery.com/tech/titanic-wireless-120411.html. Accessed on 16 June 2012.

Daniels, R., and R. Heath. (2007). '60 GHz wireless communications: Emerging requirements and design recommendations', *IEEE Vehicular Technology Magazine* 2(3): 41–50.

Department of Telecommunications (DoT). (2008). 'Amendment to the cellular mobile telephone service licence agreement issued prior to 2001.' Available at: http://www.dot.gov.in/as/cmts_137.pdf. Accessed on 3 January 2013.

———. (2009). *Auction of 3G and BWA Spectrum: Revised Information Memorandum.* Available at: http://www.dot.gov.in/sites/default/files/3g.pdf. Accessed on 10 February 2013.

———. (2011). 'National Frequency Allocation Plan (NFAP)-Draft India remarks in the National Frequency Allocation Table.' Available at http://www.dot.gov.in. Accessed on 10 January 2012.

———. (2012). 'Auction of spectrum in 1800 MHz and 800 MHz band: The information memorandum.' Available at: http://www.dot.gov.in. Accessed on 30 August 2012.

Desai, A. (2006). *India's Telecommunications Industry: History, Analysis, Diagnosis.* New Delhi: Sage Publications.

Doyle, Chris. (2004). 'The economics of pricing radio spectrum.' Department of Economics, University of Warwick 16 (2004).

Economides, N., and J. Tag. (2012). 'Net neutrality on the Internet: A two sided market analysis', *Information Economics and Policy* 24(2): 91–104.

Ericsson. (2013). *Ericsson Mobility Report: On the Pulse of the Networked Society.* Available at: http://www.ericsson.com/mobility-report. Accessed on 13 November 2013.

Farley, T. (2005). 'Mobile telephone history', *Telektronikk*, 22–34. Accessed on 15 October 2013

Farrell, J., and P.J. Weiser. (2003). 'Modularity, vertical integration, and open access policies: Towards a convergence of antitrust and regulation in the Internet age.' *Harvard Journal on Law & Technology* 17: 85–134.

FCC. (2000) Procedures implementing package bidding for auction no. 31. Available at: http://www.spectrum-exchange.com/combinatorial-bidding. htm. Accessed on 25 February 2013.

———. (2004). 'A short history of radio.' Available at http://transition. fcc.gov/omd/history/radio/documents/short_history.pdf. Accessed on 10 July 2013.

———. (2010). 'Report and order: In the matter of preserving the open Internet broadband industry practices. FCC 10–201.' Available at: http://hraunfoss.fcc.gov/edocs_public/attachmatch/FCC-10-201A1. pdfS. Accessed on 10 October 2013.

———. (2011a). 'AT&T and T-Mobile–WT Docket No. 11–65.' Available at: http://transition.fcc.gov/transaction/att-tmobile.html. Accessed on 15 May 2012.

———. (2011b). 'Preserving the open Internet.' Available at: http://www. gpo.gov/fdsys/pkg/FR-2011-09-23/html/2011-24259.htm

———. (2012). 'FCC online table of frequency allocations.' Available at: http://transition.fcc.gov/oet/spectrum/table/fcctable.pdf. Accessed on 8 December 2012.

Gandal, N., D. Salant, and L. Waverman. (2003). 'Standards in wireless telephone networks.' *Telecommunications Policy* 27(5): 325–32.

Garber, L. (2002). 'Will 3G be the next big wireless technology', *IEEE Computer* 35(1): 26–32.

Global mobile Suppliers Association (GSA). (2013). 'LTE ecosystem wall chart.' Available at: www.gsacom.com. Accessed on 10 April 2013.

Gruber, Harald, and Frank Verboven. (2001). 'The evolution of markets under entry and standards regulation—the case of global mobile tele-communications', *International Journal of Industrial Organization* 19(7): 1189–212.

Hazlett, T. (1990). 'The rationality of U.S. regulation of the broadcast spectrum', *Journal of Law and Economics* XXXIII(1, April): 133–74.

Hermalin, B., and M. Katz. (2007). 'The economics of product-line restric-tions with an application to the network neutrality debate', *Information Economics and Policy* 19(2): 215–48.

Hoffman, Karla. (2011). 'Spectrum auctions.' In J. ennington, E. Olinick, and D. Rajan (Eds), *Wireless Network Design*, pp. 147–76. NY: Springer.

Klemperer, Paul. (1999). 'Auction theory: A guide to the literature', *Journal of Economic Surveys* 13(3): 227–86.

———. (2002). 'What really matters in auction design', *Journal of Economic Perspectives* 16(1): 169–89.

Huawei. (2013). 'Whitepaper on spectrum.' Available at www.huawei.com/ ilink/en/download/HW_204545. Accessed on 15 June 2013.

Investment Information and Credit Rating Agency (ICRA). (2002). *Industrial Watch Series: The Indian Telecommunication Industry*. Mumbai, India: ICRA.

International Journal of Electromagnetic Compatibility (IJEC). (2013). 'U.S. Navy tests military spectrum sharing potential.' Available at: http://www.interferencetechnology.com/u-s-navy-tests-military-spectrum-sharing-potential/. Accessed on 10 August 2013.

Jakhu, Ram. (2007). 'International regulatory aspects of radio spectrum management: Implications for developing countries like India', p. 15. Available at: http://www.ictregulationtoolkit.org/Documents/ Document/Document/3301. Accessed on 15 May 2013.

Joskow, Paul L., and Roger G. Noll. (1999). 'The Bell doctrine: Applications in telecommunications, electricity, and other network industries.' *Stanford Law Review* 51(5, May): 1249–315.

Karimi, H.R.F., M. Lapierre, and E.G. Fournier. (2010). 'European harmonized technical conditions and band plans for broadband wireless access in the 790–862 MHz digital dividend spectrum', 2010 IEEE Symposium on New Frontiers in Dynamic Spectrum, 1–9, Singapore, IEEE, 6–9 April 2010.

Kishore, K., and V. Sridhar. (2013). 'A potential solution to cable digitization', 7 March, *Financial Express*.

Kramer, J., L. Wi5worra, and C. Weinhardt. (2012). 'Net neutrality: A progress report', *Telecommunications Policy*. Available at: http://dx.doi. org/10.1016/j.telpol.2012.08.005. Accessed on 24 November 2013.

Liebowitz, Stan J., and Stephen E. Margolis. (1995). 'Path dependence, lock-in, and history', *Journal of Law, Economics, & Organization* 11(1): 205–26.

Lin, P., J. Zhang, Q. Zhang, and M. Hamdi. (2013). 'Enabling the Femtocells: A cooperation framework for mobile and fixed-line operators', *IEEE Transactions on Wireless Communications* 12(1): 158–67.

Milgrom, Paul, Jonathan Levin, and Assaf Eilat. (2011). 'The case for unlicensed spectrum.' Available at SSRN 1948257.

Mirrlees, James A. (1986). 'The theory of optimal taxation', *Handbook of Mathematical Economics* 3. USA, Elsevier, pp. 1197–249.

Mitola, J., (2000). Cognitive radio: An integrated agent architecture for software defined radio, Doctoral thesis, Royal Institute of Technology (KTH), Sweden. Available at: http://web.it.kth.se/~maguire/jmitola/ Mitola_Dissertation8_Integrated.pdf. Accessed on 10 February 2013.

Mölleryd, B.G., and Markendahl, J. (2012). 'Valuation of spectrum for mobile broadband services—the case of Sweden and India.' Paper submitted to the regional ITS India Conference 2012, New Delhi, 22–24 February 2012. Available at: http://www.its2012india.com/topics/Spectrum%20 and%20Technology/ValuationofSpectrumforMobileBroadbandServices TheCaseofSwedenandIndia.pdf. Accessed on 9 November 2012.

Musacchio, J., G. Schwartz, and J. Walrand, 'Network economics: Neutrality, competition, and service differentiation.' In B. Ramamurthy, G. Rouskas, and K. Sivalingam (eds), *Next-Generation Internet Architectures and Protocols*. UK: Cambridge University Press.

Musgrave, Richard A., and Peggy B. Musgrave. (2011). *Public finance in theory and practice*. Fifth Edition, New Delhi, India: Tata McGraw-Hill.

National Association of Software and Services Companies (NASSCOM). (2010). *Global ER&D: Accelerating Engineering with Indian Innovation*. New Delhi.

Naone, E. (Aug 18, 2009). Wi-Fi via White Spaces, *MIT Technology Review*. Available at: http://www.technologyreview.com/news/414855/wi-fi-via-white-spaces/. Accessed on 16 October 2013.

New York Times (NYT). (2013). 'Facebook leads an effort to lower barriers to Internet access', 21 August.

Njoroge, P., A. Ozdagler, N. Stier-Moses, and G. Weintraub, 'Investment in two-sided markets and the net-neutrality debate', Decision, Risk, and Operations. Working Papers Series, DRO-2010-05, Columbia Business School, July 2010. Available at: http://www4.gsb.columbia. edu/filemgr?file_id=735208. Accessed on 10 March 2014.

Nokia Siemens Networks (NSN). (2007). 'LTE and WiMax technology and performance comparison.' Available at: http://projects.comelec.enst.fr/ EW2007/Documents/Comparison_LTE_WiMax_BALL_EW2007. pdf. Accessed on 10 March 2012.

Nuechterlein, J., and P.J. Weiser. (2005). *Digital Crossroads*. Cambridge, MA, USA: MIT Press.

Ostrom, Elinor. (1990). 'Governing the commons: The evolution of institutions for collective action.' UK, and USA: Cambridge University Press.

Prasad, R. (2010). 'The value of 2g spectrum in India', *Economic and Political Weekly* 45(4): 25–8.

Prasad, R. (2011a). *Digital Crossroads*. Cambridge, MA: MIT Press.

———. (2011b). '2008 licensing policy: Conceptual issues', *Economic and Political Weekly* xlvI(53, 31 December). Available at: http://www.epw. in/commentary/2008-telecom-licensing-policy-conceptual-issues.html. Accessed on 15 January 2012.

———. (2011c). '2008 licensing policy: Conceptual issues', *Economic and Political Weekly* XLV (53).

Prasad, R., and V. Sridhar. (2008). 'Optimal number of mobile service providers in India: Trade-off between efficiency and competition', *International Journal of Business Data Communications and Networking* 4(3): 69–81.

———. (2009). 'Allocative efficiency of the mobile industry in India and its implications for spectrum policy', *Telecommunications Policy* 33(9): 521–33.

Prepaid MVNO. (2013). 'List of prepaid MVNOs.' Available at: http://www.prepaidmvno.com/mvno-companies/eu-mvno-companies/germany-mvno-companies/. Accessed on 18 June 2013.

President's Council of Advisors on Science and Technology. (2012). 'Realizing the full potential of government-held spectrum to spur economic growth.' Available at: http://www.whitehouse.gov/sites/default/files/microsites/ostp/pcast_spectrum_report_final_july_20_2012.pdf. Accessed on 7 February 2013.

Qualcomm. (2012). 'IEEE 802.11ac: The next generation of Wi-Fi standards', May. Available at: http://www.qualcomm.com/media/documents/files/ieee802-11ac-the-next-evolution-of-wi-fi.pdf. Accessed on 12 February 2013.

Rapport, T. (2002). *Wireless Communications Principles and Practice*. Prentice Hall Inc. New Jersey, USA.

Schmidt, T., and A. Townsend. (2003). 'Why Wi-Fi wants to be free? *Communications of the ACM* 46(5): 7–52.

Shanab, L.A., B. El-Darwiche, G. Hasbani, and M. Mourad. (2007). 'Telecom infrastructure sharing: Regulatory enablers and economic benefits.' Available at: http://www.booz.com. Accessed on 20 March 2013.

Shapiro, C. (2003). 'Technology cross-licensing practices: FTC versus Intel (1999).' In John E. Kwoka and Lawrence White (eds.), *The Anti-trust Revolution*, pp. 350–72. USA: Oxford University Press.

Shapiro, C., and H.R. Varian. (1999). 'Information rules. A strategic guide to the network economy. Boston: Harvard Business School Press.

Sidak, G. (2007). 'What is the network neutrality debate really about?' *International Journal of Communication* 1 (2007): 377–88.

Siegel, M. (2009). *AT&T Spokesman on Clark Howard Radio show*. Reported on CNET. Available at: http://www.cnet.com/news/is-at-t-playing-gatekeeper-to-the-wireless-web/. Accessed on 10 March 2014.

Sridhar, Kala, and V. Sridhar. (2010). 'One paisa revolution', 4 August, *Financial Express*.

Sridhar, V. (2006a). 'A battle of widths', 6 July, *Financial Express*.

Sridhar, V. (2006b). 'How much do you pay as telecom taxes?' 25 August, *Financial Express*.

———. (2006c). 'Mega telecom partnerships', 12 September, *Financial Express*.

———. (2012a). *The Telecom Revolution: Technology, Regulation and Policy*. New Delhi, India: Oxford University Press India.

———. (2012b). 'Untangling the spectrum pricing knot', 20 October, *Financial Express*.

———. (2013). 'Maximizing spectrum revenue', 5 March, *Business Standard*.

Sridhar, V., and C. Buchi. (2012). 'Is Wi-Fi offloading really good?' *Voice & Data* 82–3 (September).

Sridhar, V., and D. Manjunath. (2012). 'For a dynamic spectrum regime', 17 July, *Economic Times*.

Sridhar, V., and G. Venkatesh. (2009). 'Let the traffic flow', 14 September, *Business Line*.

Sridhar, V., and R. Prasad. (2011). 'Towards a new policy framework for spectrum management in India', *Telecommunications Policy* 35(2): 172–84.

Sridhar, V., T. Casey, and H. Hämmäinen. (2012). 'Systems dynamics approach to analyzing spectrum management policies for mobile broadband services in India', *International Journal of Business Data Communications and Networking* 8(1): 37–55.

Sridhar, V., T. Casey, and H. Hämmäinen. (2013). 'Flexible spectrum management for mobile broadband services: How does it vary across advanced and emerging markets? *Telecommunications Policy* (37, Special Issue on Cognitive Radio): 178–91.

Shannon, C.E. (1948). 'A mathematical theory of communication', *Bell System Technical Journal* 27 (July–October): 379–423, 623–6.

Stephenson. M. (2002). *The Marconi Wireless Installation in R.M.S. Titanic*. Available at: http://marconigraph.com/titanic/wireless/mgy_wireless.html. Accessed on 13 June 2012.

Sum, C., G. Villardi, Z. Lan, C. Sun, Y. Alemseged, H.N. Tran, J. Wang, and H. Harada. (2011). 'Enabling technologies for a practical wireless communication system operating in TV white space', *ISRN Communications and Networking* 13 (doi:10.5402/2011/147089).

Swamy, S. (2011). *2G Spectrum Scam*. New Delhi, India: Har-Anand Publications Pvt. Ltd.

Tanenbaum, A. (1996). *Computer Networks*. NJ, USA: Prentice Hall.

Taparia, A., T. Casey, and H. Hammainen. (2012). 'Towards a market mechanism for heterogeneous secondary spectrum usage: An evolutionary approach', IEEE International Symposium on Dynamic Spectrum Access

Networks (DYSPAN), 16–19 October 2012, Bellevue, Washington, USA, 142–53.

Telecommunications Regulatory Authority of India (TRAI). (2005). 'Recommendations on spectrum related issues.' Available at Telecom Regulatory Authority of India website: http://www.trai.gov.in/

———. (2006). 'TRAI issues amendment to interconnect usage charges regulation. Available at http://www.trai.gov.in on. Accessed on 29 April 2006.

———. (2007). 'Draft recommendations on growth of broadband.' Available at: http://www.trai.gov.in on. Accessed on 20 September 2007.

———. (2008a). 'Recommendations on issues relating to mobile television service.' Available at: http://www.trai.gov.in/WriteReadData/Recommendation/Documents/FINALRECOMENDATIONS.pdf. Accessed on 23 April 2008.

———. (2008b). 'Consultation paper on issues relating to 3rd phase of private FM radio broadcasting', January. Available at: http://www.trai.gov.in. Accessed on: 15 January 2011.

———. (11 May 2010). *Spectrum Management and Licensing Framework.* Available at: www.trai.gov.in. Accessed on 25 July 2011.

———. (2012a). 'Recommendations on auction of spectrum.' Available at: http://www.trai.gov.in

———. (2012b). 'Recommendations on "Prescribing minimal channel spacing within licensed service area in FM Radio sector in India".' Available at: www.trai.gov.in

———. (2013a). 'Consultation paper on valuation and reserve price of spectrum.' Available at: http://www.trai.gov.in

———. (2013b). 'Recommendations on IMT-advanced mobile wireless broadband services.' Available at: www.trai.gov.in

———. (2011). *Report on the 2010 value of spectrum in the 1800 MHz band.* Available at: http://www.trai.gov.in/Content/Recommendation Description.aspx?RECOMEND_ID=231&qid=0. Accessed on 23 April 2012.

Varian, Hal R. (1992). *Microeconomic Analysis*, Vol. 2. NY: Norton.

Vascellaro, Jessica E. (2012). 'Audit Faults Apple Supplier', *Wall Street Journal*, 30 March. Available at: http://online.wsj.com/article/SB10001424052702303404704577311943943416560.html. Accessed on 15 May 2013.

Vickrey, William. (1961). 'Counterspeculation, auctions, and competitive sealed tenders', *Journal of Finance* 16(1): 8–37.

Werbach, K. (2009). 'The wasteland: Anticommons, white spaces and the fallacy of spectrum', *Arizona Law Review* 53(1): 213–54.

WPC. (2008). 'Subscriber base criterion for allotment of GSM spectrum.' Available at: http://www.wpc.dot.gov.inS. Accessed on 8 December 2013.

WPC. (2011). 'Draft India remarks in the national frequency allocation table.' Available at: http://www.dot.gov.in

Wyatt, Edward. (2013). 'FCC move to ease wireless congestion', *New York Times*, 20 February.

Xavier, P., and D. Ypsilanti. (2006). Policy issues in spectrum trading. *Info* 8(2), 34–61, doi: 10.1108/14636690610653581.

Index

access: fee 176, 177, 178; network 40, 171, 198, 203, *see also* Radio Access Network (RAN), Wi-Fi access network; service 151, 172, 202, 230, 231, 233, 270, 275, 294, 345; spectrum 154, 286, 295

acquisitions: inter-region 192–4; intra-region 194–8

administration costs for operators 109–11

administered pricing, Smith-NERA method of 74

Advanced Mobile Phone Service (AMPS) 21, 24

advertising revenue 177, 317

Aggarwal, Varun 3, 4

Aircel 194–5, 198–9, 203

Airtel 106, 192, 193, 198, 199, 201, 202, 215, 217, 237, 238, 249, 253, 264, 265, 274, 278, 296

airwaves 1; *see also* public airwaves

All India Radio 303

AM: 11–12, 301, 303; broadcasting 346

Ambani, Anil 199, 233

Ambani, Mukesh 199, 249

Amplitude Modulation *see* AM

AMPS *see* Advanced Mobile Phone Service (AMPS)

Android 181, 213

antenna 3, 18, 212, 308, 346

anti-trust activity 182

Apple 149, 181, 183, 184, 213, 214, 215, 216

ARPU (average revenue per user), decline in 237

ascending auctions *see* auctions

auctions: ascending, 90, 92, 96, 97, 98, 102, 227, 241, 252, 261, 338; clock 88, 339; common value 93, 338; descending 87, 88, 338; Dutch 88, 338; dynamic 87, 90, 94, 338; English 88, 339; first-price 339; Japanese 88, 339; multi-stage 294; multi-unit 95–6, 339; package 89, 98–102, 339; private value 92, 93, 339; reverse 316, 317, 318; sealed-bid 87, 88, 90, 340; second-price 88, 92, 98, 339; sequential 96; simultaneous ascending 96–8, 339; static 88, 340; Vickrey 88, 340

Augere 243, 247

Australia 24, 138, 139, 140, 141, 201, 203, 314

Bangalore 212, 292
'beauty contests' 85, 131, 338
'beauty parade' method 222
Bell System 16, 18, 21, 167
Berners-lee, Tim 166
Bit Torrent 172
Bluetooth 10, 14, 34, 42, 145, 146
Bose, Jagadish Chandra 3, 8
Brazil 35, 209, 313
Britain 22
Broadband Wireless Access *see* BWA
Bullington, Kenneth 18
BWA: 34, 103, 122, 131, 132, 133,
    195, 239, 241, 242, 243, 244,
    249, 278, 295, 304, 315, 341;
    spectrum auction financials 247–8

Canada 24, 201, 323
canonical auction format 88–9, 90–1
capacity-constrained areas 18–19
capacity-enhancing technological
    progress 60–2
carrier aggregation: 30; use of, by
    5G technologies 275
Category: A 105, 108, 110, 113,
    114, 115, 223, 226, 227, 244,
    261, 266, 286, 287, 288, 304,
    340; B 105, 108, 116, 223, 227,
    244, 261, 266, 286, 287, 288,
    304, 340; C 105, 108, 116, 222,
    223, 226, 228, 244, 286, 287,
    288, 340
C-band 318
CDMA 2000 28
CDMA: 18, 23, 25–7, 28, 30, 32,
    33, 34, 35, 36, 76, 77, 78, 108,
    136, 207, 214, 222, 230, 231,
    232, 233, 234, 236, 251, 252,
    253, 255, 270, 275, 289, 290,
    295, 296, 331, 332, 334, 335,

336, 346; EVDO 37, 138, 331,
    332
CEPT (Confederation of European
    Postal and Telecommunications)
    9, 23
Chicago School 182
Chidambaram, P. 265
China 24, 30, 31, 36, 37, 77, 249,
    313
Cingular 195
Cisco 168, 169, 206, 207, 216
Clearwire 194, 210
clock auctions *see* auctions
CMSP (Cellular Mobile Service
    Provider) 226, 345
CMTS (Cellular Mobile
    Telecommunication Service)
    241, 242, 244; *see also* CMSP
Coase Theorem 58, 60, 130, 336
Cobb-Douglas production function
    111, 340
Code Division Multiple Access
    see CDMA
Cognitive Radio (CR): 43, 153,
    155, 156, 157, 158, 327, 332;
    possibilities of 289–91
coherer 3, 4
collusion 94–5, 338
combinatorial auctions 89, 99
Comcast 128, 196
common goods 54, 336
common property resources 336
common value auctions
    see auctions
commons 8, 37, 40, 53, 54, 55, 57,
    58, 59, 60, 61, 62, 63, 64, 65, 66,
    67, 68, 69, 71, 72, 135, 144, 145,
    147, 149, 151, 153, 155, 157,
    159, 160, 161, 177, 188, 197,
    207, 208, 215, 301, 322, 336, 337

Comptroller and Auditor General (CAG) 239, 267
congestion in networks 40, 57, 60, 62, 167, 168–73, 185, 186, 337
context-aware mobile applications 327
contracts 75, 289
convergence 181, 212, 323
co-opetition 213, 218
corDECT 20, 232
cordless 2, 10, 19, 20, 38, 145
coverage-constrained circles 18, 135, 197
Cramton, Peter 98
Crocioni, P. 127, 138
cross-group externalities see externalities
cross-holdings 181

data-oriented 3G spectrum 266
decimeter band see UHF
decimeter wave see UHF
descending auctions see auctions
Digital Dividend Spectrum 311–16
Digital Living Network Alliance see DLNA
digital: broadcasting, 7, 13, 313; cellular, 32, 75, 76, 223, 333; living, 42, 146; signals, 23, 307, 315; standard, 23, 76; switchover 312, 313, 315; technology 22, 23, 76, 79; transmission 312, 315; tuner 307, 315, 319; TV 308, 310, 312
DirecTV 319
'dirty road' fallacy 179, 180
DLNA 42, 146, 343
DoCoMo 192, 193, 204, 210, 211
Doordarshan 302, 309, 310, 311, 312, 322

DSL (Digital Subscriber Line) 43, 136, 151, 152
DTH (Direct-To-Home) 125, 192, 311, 318–21, 346
DTH TV, adoption rate of 319
Dual Technology Licensing 233–4
Dutch auctions see auctions
dynamic auctions see auctions
Dynamic Spectrum Management (DSM) 43

ECP (End-user Connectivity Provider) 165, 166, 167, 168, 173, 174, 175, 176, 177, 178, 179, 180, 183, 188
electromagnetic spectrum 1
electromagnetic waves 2
English auctions see auctions
Ericsson 21, 215, 327
'Erlangs' 235
Essar 199
ETSI see European Telecommunications Standards Institute
Europe 17, 20, 21, 22, 23, 24, 30, 37, 75, 76, 77, 125, 129, 137, 172, 204, 208, 222, 288
European Telecommunications Standards Institute (ETSI) 16, 23, 28, 44, 333
excludable goods 53, 54, 55, 166, 325, 336
externalities: 336; cross-group 176; network externalities 338

Facebook 164, 181, 213, 216, 327, 329
FCC (Federal Communication Commission) 9–10, 12, 17, 26,

38, 40, 41, 85, 90, 99, 102, 135,
145, 158, 159, 165, 184, 185,
195, 196, 301, 309, 311, 315,
316, 317, 318, 322, 332
FDD 35, 46, 47, 48, 313, 314
FDI (Foreign Direct Investment)
237, 286
FDMA 25, 33, 333
Federal 5, 9, 17, 85, 102, 122, 144,
145, 160, 161, 183, 301
Federal Communication
Commission see FCC
Federal Radio Commission,
USA 5
Femtocell: 44, 135–7, 162,
200, 341; architecture 136;
partnerships 204–7
first-price auctions see auctions
flat termination fee 177–8
flexible spectrum 199
flexible use 119, 126, 138
FM: 12; auction 304; broadcasting
346
4G 29, 332; see also LTE
France 35, 82, 124, 125, 205, 206,
208, 209, 288, 313
free-rider 55
frequency allocation, complexity
of 11
Frequency Division-Long Term
Evolution (FD-LTE) 30, 77,
280, 289
Frequency Modulation see FM
FTA (Free-to-Air) 311

Germany 2, 21, 35, 204, 206, 209,
288, 313
goods, economic classification of
53–5
GPRS (General Packet Radio
Service) 27, 32, 34

GPS (Global Positioning System)
10, 195
Groupe Spécial Mobile see GSM
Gruber, Harald 76, 79, 81
GSM Association 333
GSM: 9, 15, 23–5, 26, 27, 28, 30,
32, 34, 35, 36, 37, 44, 76, 77,
108, 113, 136, 139, 204, 216,
222, 226, 228, 231, 232, 233,
234, 236, 239, 251, 252, 255,
270, 274, 278, 280, 289, 290,
295, 296, 331, 333; coordinated
adoption of 24; disadvantages
of 27
GSMA see GSM Association

HCP (Hosting Connectivity
Provider) 174, 175
Hetnets 42, 43, 44
Hexacom 192
HSDPA 28, 333
HSPA 28, 29, 34, 36, 77, 280, 333
HSUPA 28, 333
Hutchison 193

ICT 7, 63, 68, 124, 141, 142, 180,
188, 213, 326
IEEE 14, 29, 30, 31, 34, 38, 39,
40, 41, 42, 44, 147, 149, 151,
152, 157, 158, 171, 218, 334
IMT2000 27–8, 331, 333
IMT-MC/DS 334
In-Building Solutions (IBS) 42
INSAT 13, 122, 318, 319, 320,
321, 322, 342
International Mobile
Telecommunications (IMT) 27
International Telecommunications
Union (ITU) 5, 7, 9, 11, 14,
15, 27, 43, 71, 72, 124, 220,
222, 250, 296, 313, 334;

International Radio Regulations of 9

International Telegraph Union *see* International Telecommunications Union (ITU)

inter-region acquisitions *see* acquisitions: inter-region

inter-region *see* leasing of spectrum: inter-region

intra-region acquisitions *see* acquisitions: intra-region

intra-region leasing *see* leasing of spectrum: intra-region

iPhone 6, 151, 172, 181, 183

ITU *see* International Telecommunications Union

ITU-Telecom Standardization Sector 15–16

Japanese auctions *see* auctions

Ka-band 300

Kennard, William 322

Kindle 204

Korea Mobile Telecommunications (KMT) Company 22

Ku-band 300

landline 19, 20, 38, 43, 121, 150, 152, 171, 185, 241

L-band 37, 194, 195, 300

leasing of spectrum: spectrum manager model 130; intra-region leasing 131–4; inter-region leasing 131–4; between MNOs and MVNOs/ISPs 134

liberalization 118, 138, 142, 220, 239, 249, 251, 253, 254, 266, 270, 303, 306

licensed spectrum 65–6

licences: migration of 251–3; renewal of 253–5; quashed 194, 219, 239, 241, 255, 265, 267, 271, 295, 323

Li-Fi *see* Variable Light Communications (VLC)

Lightsquared 194, 195

Long Term Evolution *see* LTE

'Loon' project 212

LSA (Licensed Service Area) 121, 132, 134, 154, 157, 201, 202, 203, 226, 229, 240, 242, 244, 245, 246, 247, 248, 263, 264, 268, 276, 279, 281, 283, 286, 287, 294, 341

LTE 28, 30, 34, 37, 44, 128, 138, 152, 192, 243, 249, 257, 325, 327, 332, 334

macro-cell 148, 158, 160, 204

managed commons 337

Marconi, Guglielmo 3, 4, 6, 7, 8, 16

marginal revenue 337

marginal: cost, 57, 58, 78, 79, 186, 337; productivity, 113; revenue 58, 59, 61

m-business 29

m-commerce 29

Mercury autocoherer 3

mergers and acquisitions 134–5

micro-cell 148, 160, 299

Microsoft 213, 215, 216, 321

Microsoft-Skype 151

microwave frequencies 14

Mi-Fi 343

Minow, Newton 309

mobile sector, important milestones in 294–6

mobile technologies, comparison of 32–3

Mobile Virtual Network Operator
(MVNO) 134, 203–4, 205, 206,
210, 316, 342
mobile number portability 79, 338
modelling auctions 90–1
modulation schemes 41, 78
monopolization of content 305
Morse code 3, 7
Motorola Mobility 213, 215
multimedia 165
multi-SIM mobile handsets 156–7,
201
multi-stage auctions *see* auctions
multi-unit auctions *see* auctions
MVNO *see* Mobile Virtual
Network Operator

narrow-band approach 160
National Frequency Allocation Plan
(NFAP) 10, 11, 12, 20, 49, 124,
220, 292, 293, 303, 312, 313,
321, 342; *see also* WPC
net-neutrality 189, 344
network externalities *see* externalities
New Telecom Policy, 2012
249–50
New Zealand: auctions in 89, 90,
141; property rights in 138
NFAP *see* National Frequency
Allocation Plan (NFAP)
non-excludable goods 53
non-rivalrous goods 53
NTP (National Telecom Policy),
200, 270, 289–90, 322

OfCom 129, 137, 158
OFDMA 334
open access 64, 207, 316, 337
open architecture 182, 213
open ascending-bid auctions 88
open descending-bid auctions
see auctions: Dutch

opportunity cost 112, 116, 123, 252
Optus 204, 205
Orange 196, 204, 205, 208
Orthogonal Frequency Division
Multiple Access (OFDMA) 334
Ostrom, Elinor 54, 63–4, 67
OTT (Over-The-Top) 151, 343;
see also VoIP
outsourcing 291
over-the-air broadcasting 125, 310

package auctions *see* auctions
Paris Wi-Fi 147
patents 65, 215, 216
peer-to-peer applications 169, 172–3
pico-cell 207
Post Telegraph and Telephones
(PTTs) 17
Prasar Bharati 302, 315, 322;
see also Doordarshan
private commons 337
private goods 54, 337
private value auctions *see* auctions
public airwaves 309–11
public: goods, 54, 55, 122; policy,
60, 62, 69, 142; sector, 106, 113,
118, 121, 122–3, 124, 314, 326
pure public goods 55, 337

Qatar 204
Qualcomm 26, 30, 77, 128, 207,
234, 243, 249, 327
quashed licences *see* licences: quashed
Quippo Towers 198, 199

Radio Access Network (RAN) 198,
203
Radio Frequency Identification
(RFID) 146, 343
radio frequency spectrum 1
radio spectrum: 1–2, 65, 300–1;
first auction of 89; policy 306

Radiolinja 24
Reliance Jio 199, 275, 278, 280
reserve price 95, 99, 103, 254, 257,
   259, 261, 266, 267, 270, 276,
   278, 279, 280, 296, 304, 306, 316
reserve price 339
RFID *see* Radio Frequency
   Identification
rivalrous goods 53, 337
roaming 9, 23, 24, 26, 29, 37, 42,
   75, 118, 131, 138, 157, 162, 200,
   201, 202, 203, 209, 249, 289
roll-out obligations 79, 119, 120,
   129, 154, 229, 244, 249

Samsung 181, 214, 327
SATCOM 320
Satellite Television 318–21
scam 239
Scandinavia 75
Scotland 206
sealed bid auctions *see* auctions
second-price auctions *see* auctions
security 145, 149, 150, 329
sequential auctions *see* auctions
SIM 24, 150, 157, 290, 344
SIM-based authentication 150
simultaneous ascending auctions
   see auctions
Singapore 36, 204, 209
single-homing 176, 186
Singtel 204
Sistema 193, 240, 265, 267, 268, 269
Skycell 192, 193
Skype 66, 151, 183, 188, 215
small-cell 206, 207
smartphones 40, 66, 149, 152,
   212, 249
SMS 23
Softbank 192
Sony 213
Sotheby 85

South Korea 26
spectrum management, US
   administration's approach to 26
spectrum management regimes:
   command and control 71; 74;
   command and control vs flexible
   use 119–21; flexible use 71;
   spectrum commons 71
spectrum: allocation 5, 8, 9,
   71–2, 113, 122, 124, 131, 132,
   215, 220, 231, 232, 233, 241,
   251, 252, 254, 271, 280, 295,
   see also Subscriber Based Criteria
   (SBC) for spectrum allocation;
   assignment 72, 86, 119, 155, 226,
   229, 270, 272, 281, 283, 284;
   auctions *see* auctions; availability
   16, 232, 280, 326; brokers 290;
   capacity 22, 128; commons 62, 64,
   66–7, 69, 72, 144, 145, 147, 149,
   151, 153, 155, 157, 159, 161,
   188, 208; digital dividend 124,
   126–6, 142, 289, 307, 311–16,
   323; economic classification
   of 55–7; exchange 102, 351;
   fragmentation 219, 223, 227, 229,
   231, 233, 235, 237, 239, 241,
   243, 249, 251, 253, 255, 257,
   259, 261, 265, 267, 271, 275,
   281, 289, 291, 293, 326; HHI
   284, 285; holding, 65, 86, 128,
   136, 154, 194, 197, 229, 251,
   252, 254, 266, 278, 283, 285,
   286; leasing 131; liberalization,
   138; licences, 65, 126; managers,
   9, 71, 72, 85; markets, 66, 121,
   123, 127, 128, 129, 203, 327;
   owners, 65, 120, 289; policy
   68–86, 293, 306; pooling 153,
   250; price 81–3, 238, 288, 302;
   reallocation of 66, 124; refarming
   26, 37, 139–41, 142, 254, 256,

258, 266, 275; regime, 120, 140, 142, 199, 239, 289; regulation 4, 185; reuse 198; rights 102, 104, 138, 162, 199; sale 130; scam 239, 356; sensing 147; sharing 67, 130, 147, 148, 153, 154, 160, 161, 162, 186, 203, 286, 326, 329; technology 38, 69, 149; trading 66, 127–8, 129, 139, 141, 188, 281, 352, 353; transfer 129, 130; transitions 119, 121, 123, 125, 127, 129, 131, 135, 137, 139, 141; usage 35, 36, 57, 69, 82, 83, 126, 128, 131, 154, 255, 275, 278, 286, 296, 316, 317; usage charge 126–7; utilization 250; value of 73; winners 243, 249
spectrum valuation: cash flow method 105–11; production function method 111–15
Spice 192, 193, 196, 240
Sprint 192, 193, 194, 195, 205
Starbucks 150, 210
start-up 111, 177, 226, 227, 228, 230, 232, 233
static auctions see auctions
Stockholm 31
streaming 28, 29, 42, 149
Subscriber Based Criteria (SBC), for spectrum allocation 228–30
Subscriber Identification Module see SIM
surcharge 180, 200
surveillance 53
Swan Telecom 194
Swisscom 209
Switzerland 7, 205, 209, 313

tablet 66, 213
takeaways 278
taxation 82, 266, 300

TDD 47, 48, 313, 314
TDM 28, 31
TDMA 25, 27, 32, 33, 56, 335
TD-SCDMA 31, 34, 37, 335
technological advances 60, 61, 77, 145
Telecom Regulatory Authority of India (TRAI) 104, 105, 107, 109, 114, 116, 126, 154, 200, 203, 225, 229, 230, 232, 233, 236, 239, 241, 254, 255, 257, 259, 260, 265, 270, 273, 275, 294, 295, 296, 302, 303, 305, 314, 315, 345
telecommunity 312
teledensity 235
Telefonica 209
telegraph 7
Telenor 193, 194, 237, 265, 276, 277
Telewings 193, 263, 264, 265
termination fee 176; see also flat termination fee, tiered termination fee
Terrestrial TV 309–11
tertiary 69, 148
theorists 89, 90
Third Generation Partnership Project (3GPP) 16, 331, 332
3G: 331; roaming pact 201–3; spectrum auction financials 245–6; technologies 28
tiered termination fee 177–80
Tikona 243, 247, 248
Time Division Multiple Access see TDMA
Time Division Duplexing (TD-LTE) 30
Time Division-Synchronous Code Division Multiple Access see TD-SCDMA

Titanic 4, 6, 7, 8, 49, 55, 301
T-Mobile 128, 195, 196, 204
toll goods 55, 337
Townsend 50
tragedy of the anti-commons 65, 337
tragedy of the commons: 57–8, 63,
    177, 301, 337; resolving 58–60
TRAI *see* Telecom Regulatory
    Authority of India
treasury 226
trunking efficiency 78, 87, 153
TV broadcasts, models of 308
TV White Space 157–8, 321;
    see also digital: dividend
TVWS *see* TV White Space
2G 331
2G radio systems 22
2G technologies 27
two-sided markets 173, 174, 175,
    177, 184, 186

UASL 202, 230–3, 294, 295, 345
Ubiqusys 206
Uganda 193
UHF 335
Ultra High Frequency *see* UHF
Ultra Wide Band (UWB) 41
UMTS *see* Universal Mobile
    Telecommunications Services
Unified Access Service Licence
    see UASL
Unitech 194, 237, 265
Unitech Wireless, acquisition of,
    by Telenor 237
Universal Mobile
    Telecommunications Services
    (UMTS) 28, 34, 288, 335
Universal Service Obligation
    (USO) 141–2, 342
unlicensed spectrum, adoption of 66
US Navy 327

USA 128, 188, 209, 325
usage-based roaming surcharge 200
USO *see* Universal Service
    Obligation

Variable Light Communications
    (VLC) 45
vehicular networks 34, 41, 146
Venezuela 313
venture 128, 198, 205, 213, 267
Verizon 128, 165, 196, 208
Very High Frequency *see* VHF
VHF 335
Viacom 309
Vickrey auctions *see* auctions
Vickrey's truth serum 92–3
Videocon 240, 263, 264, 265
videoconferencing 56
video-on-demand 212
Vodafone 106, 198, 201, 204, 253,
    265, 274, 278, 280, 282
Voice Over Internet Protocol
    see VoIP
VoIP 151, 169, 170, 183, 184,
    215, 336
Vonage 183

Washington 210
waterbed effect 176
WCDMA 28, 30, 32, 34, 35, 36,
    37, 44, 77, 136, 207, 242, 259,
    280, 289, 336
Whitehouse 70
wideband 28, 56, 78
Wideband Code Division Multiple
    Access *see* WCDMA
Wi-Fi: 10, 34, 37, 38, 39, 40, 41,
    42, 43, 44, 57, 66, 69, 136, 137,
    145, 146, 147, 148–53, 156,
    157, 158, 162, 172, 183, 206,
    208, 209, 210, 211, 212, 215,

216, 217, 321, 323; community
Wi-Fi 149, 210–12; Wi-Fi
hotspots 149–51
Wi-Fi access network 38
Wi-Fi technologies, evolution
of 39
WiGig 146
WiMAX 29, 30, 31, 34, 35, 36,
43, 77, 122, 136, 171, 217,
242, 342
winner's curse 91, 93, 94, 226, 228,
289, 340
Wireless Fidelity see Wi-Fi
Wireless Local Area Network
see WLAN 37, 146, 343

Wireless Personal Area Network
see WPAN
wireless-local-loop 90
WLAN 37, 146, 343
WPAN 146
Worldwide Interoperability for
Microwave Access see WiMAX
WPC 10, 220, 231, 292, 303, 313

Yahoo! 166, 210

Zain Telecom, acquisition of, by
Bharti Airtel 192
zero-price rule 174
Zigbee 42

About the Authors

Rohit Prasad is Associate Professor of Economics at Management Development Institute (MDI), Gurgaon, India. He has a PhD in Economic Theory from State University of New York at Stony Brook, USA. After his PhD, he worked in the software industry in the USA and India in senior management positions before joining MDI, Gurgaon. His research papers have been published at leading international journals and his articles appear regularly in financial dailies, such as the *Economic Times* and *Mint*. He has been a member of the Subodh Kumar Committee on Spectrum Allocation at the DoT, and of the expert panel of the TRAI on Spectrum Pricing. His popular book titled *Start-Up Sutra* (on entrepreneurship) was published in 2013.

V. Sridhar is Professor, Centre for IT and Public Policy, International Institute of Information Technology Bangalore, India. Formerly, he was a Research Fellow at Sasken Communication Technologies, Bangalore, where most of the research for this work was carried out. Sridhar has a PhD from The University of Iowa, USA; PGDIE from National Institute of Industrial Engineering (NITIE), Mumbai; and B.E. from University of Madras. He has published many articles in leading telecom journals and about 175 articles in newspapers. His book titled *The Telecom Revolution in India: Technology, Regulation, and Policy* was published by Oxford University Press in 2012. He has also co-edited three other books. He has taught at many institutions in the USA, New Zealand, and India and has been a member of government committees in telecom and IT.